T0177440

SCUOLA NORMALE SUPERIORE

QUADERNI

A. ANDREOTTI
M. NACINOVICH

Analytic Convexity and the Principle of Phragmén-Lindelöf

PISA - 1980

ISBN: 978-88-7642-243-0

INTRODUCTION

The theory of analytic convexity originated from two very pertinent remarks of De Giorgi, Piccinini and Cattabriga.

The first remark is the following (De Giorgi and Piccinini [6], [18]).

We consider in $\mathbb{R}^3$, where x, y, t are cartesian coordinates, the differential equation

$$\frac{\partial^2 u}{\partial x^2} + \frac{\partial^2 u}{\partial y^2} = f$$

and for f real analytic in $\mathbb{R}^3$ we seek a solution u also real analytic in $\mathbb{R}^3$. There exists f real analytic in $\mathbb{R}^3$ such that the above equation does not admit any real analytic solution u in the whole of $\mathbb{R}^3$.

The second remark is the following (De Giorgi and Cattabriga [7], [8]).

We consider in $\mathbb{R}^2$, where x, y are cartesian coordinates, a differential operator with constant coefficients $P(\partial/\partial x, \partial/\partial y)$, where $P(\xi, \eta)$ is a polynomial. Then for every f real analytic in $\mathbb{R}^2$ the equation

$$P\left(\frac{\partial}{\partial x}, \frac{\partial}{\partial y}\right) u = f$$

always admits a real analytic solution u in $\mathbb{R}^2$. Thus to encounter examples as mentioned in the first remark the number of variables needs to be $\geqslant 3$.

In a subsequent paper Hörmander [15] has considered in general the following problem.

We consider in $\mathbb{R}^n$ a differential operator $P(D)$ with constant coefficients, $D = (\partial/\partial x_1, \ldots, \partial/\partial x_n)$, $P(\xi)$ for $\xi = (\xi_1, \ldots, \xi_n)$ a polynomial, $x_1, \ldots, x_n$ cartesian coordinates in $\mathbb{R}^n$. Let ω denote an open set in $\mathbb{R}^n$ and let $\mathcal{A}(\omega)$ designate the space of real analytic functions on ω. We consider the equation

$$(*) \qquad P(D) u = f$$

for $f \in \mathcal{A}(\omega)$ and we look for a solution u also in $\mathcal{A}(\omega)$.

Let P^0 be the principal part of the polynomial P and let

$$W = \{\xi \in \mathbb{C}^n | P^0(\xi) = 0\}$$

denote the asymptotic cone of the characteristic variety $\{\xi \in \mathbb{C}^n | P(\xi) = 0\}$.

For K compact in $\mathbb{R}^n$ we set

$$H_K(\xi) = \sup_{z \in K} \operatorname{Re}\left(\sum_1^n \xi_i z_i\right).$$

The result of Hörmander is the following theorem.

Let ω be open and convex. The necessary and sufficient condition for the equation $(*)$ to have a solution $u \in \mathcal{A}(\omega)$ for every choice of $f \in \mathcal{A}(\omega)$ is that the following condition holds: « given a compact convex set $K \subset \omega$ we can find another compact convex set K' with $K \subset K' \subset \omega$ and a $\delta > 0$ such that: for any plurisubharmonic function φ defined on $\mathbb{C}^n$ and satisfying the conditions

$$\begin{cases} \varphi(\xi) \leqslant H_K(\xi) + \delta|\xi| & \forall \xi \in W \\ \varphi(\xi) \leqslant 0 & \forall \xi \in W \cap i\mathbb{R}^n \end{cases}$$

we also have

$$\varphi(\xi) \leqslant H_{K'}(\xi) \qquad \forall \xi \in W \text{ ».}$$

We recall one of the forms of the Phragmén-Lindelöf principle in $\mathbb{C}^n$:

« if φ is any plurisubharmonic function in $\mathbb{C}^n$ which satisfies

$$\begin{cases} \varphi(\xi) \leqslant H_K(\xi) + \delta|\xi| & \forall \xi \in \mathbb{C}^n \\ \varphi(\xi) \leqslant 0 & \forall \xi \in i\mathbb{R}^n \end{cases}$$

then φ satisfies also the more stringent inequality

$$\varphi(\xi) \leqslant H_K(\xi) + \delta|\operatorname{Re} \xi| \qquad \forall \xi \in \mathbb{C}^n \text{ ».}$$

Because of the analogy with the statement in the theorem of Hörmander that statement is called the Phragmén-Lindelöf principle on the asymptotic variety W.

Note that if $K(\delta) = K'$ denotes the δ-neighborhood of K in $\mathbb{R}^n$ we do have

$$H_{K(\delta)}(\xi) = H_K(\xi) + \delta|\operatorname{Re} \xi|.$$

From this theorem of Hörmander one derives the fact that if a cone W admits a Phragmén-Lindelöf principle then at each one of its real points $a \neq 0$ the real part of W is pure dimensional of dimension $n-1$. This explains the first example of De Giorgi and Piccinini. Secondly for the remark of De Giorgi and Cattabriga the cone W in that case reduces to a set of complex lines issued from the origin. The Phragmén-Lindelöf principle for W is reduced then to the classical one for $\mathbb{C}$.

In this paper we consider instead of the equation $(*)$ a general Hilbert complex of differential operators with constant coefficients in $\mathbb{R}^n$

$$(1) \qquad \mathcal{E}^{p_0}(\Omega) \xrightarrow{A_0(D)} \mathcal{E}^{p_1}(\Omega) \xrightarrow{A_1(D)} \mathcal{E}^{p_2}(\Omega) \longrightarrow \dots$$

(where $\mathcal{E}(\Omega) = C^\infty(\Omega)$ and Ω is open in $\mathbb{R}^n$) obtained via Fourier (or Laplace) transform from a Hilbert resolution of a finitely generated $\mathcal{P}$-module N where $\mathcal{P} = \mathbb{C}[\xi_1, \dots, \xi_n]$ is the ring of polynomials in n-variables $\xi_1, \dots, \xi_n$ (cf. [2]);

$$(2) \qquad \dots \longrightarrow \mathcal{P}^{p_2} \xrightarrow{{}^tA_1(\xi)} \mathcal{P}^{p_1} \xrightarrow{{}^tA_0(\xi)} \mathcal{P}^{p_0} \longrightarrow N \longrightarrow 0\,.$$

Since the differential operators $A_j(D)$ have constant coefficients, we can consider aside of the complex (1) the complex

$$(3) \qquad \mathcal{A}^{p_0}(\Omega) \xrightarrow{A_0(D)} \mathcal{A}^{p_1}(\Omega) \xrightarrow{A_1(D)} \mathcal{A}^{p_2}(\Omega) \longrightarrow \dots$$

(where $\mathcal{A}(\Omega)$ is the space of real analytic functions in Ω).

Let $\mathcal{E}_0$ (resp. $\mathcal{A}_0$) denote the sheaf of germs u of C^∞ (resp. real analytic) solution of the homogeneous equation $A_0(D)u = 0$. Then we have that the cohomology of the complex (1) is given by the groups

$$H^j(\Omega, \mathcal{E}_0)$$

and the cohomology of the complex (3) is given by the groups

$$H^j(\Omega, \mathcal{A}_0)$$

(cf. [4]). For Ω open convex one has

$$H^j(\Omega, \mathcal{E}_0) = 0 \quad \forall j > 0 \qquad \text{and} \qquad H^j(\Omega, \mathcal{A}_0) = 0 \quad \forall j \geqslant 2$$

so the only interesting group is $H^1(\Omega, \mathcal{A}_0)$ which is either zero or infinite dimensional (cf. [2], [4]). The vanishing of this group is the necessary and sufficient condition that the equation

$$(**) \qquad A_0(D)u = f$$

whenever $f \in \mathcal{A}^{p_1}(\Omega)$ and satisfies the integrability conditions $A_1(D)f = 0$ admits a solution $u \in \mathcal{A}^{p_0}(\Omega)$. So this is the generalization to « overdetermined systems » of the problem considered by Hörmander.

In this paper we give, for Ω convex, the necessary and sufficient conditions for the vanishing of the group $H^1(\Omega, \mathcal{A}_0)$ in terms of the generalization of Phragmén-Lindelöf principle. This is obtained as follows.

One first considers the set of prime ideals $\mathfrak{p}_i \subset \mathcal{P}$ given by

$$\operatorname{Ass}(N) = \mathfrak{p}_1 \cup \ldots \cup \mathfrak{p}_k$$

(those are the prime ideals that are annihilators of some non zero element of N). We define the characteristic varieties

$$V_i = \{\xi \in \mathbf{C}^n | p(\xi) = 0 \ \forall p \in \mathfrak{p}_i\}$$

and the corresponding asymptotic cones V_i^0 (defined as the zero set in $\mathbf{C}^n$ of all homogeneous polynomials that appear as the principal part of some element of $\mathfrak{p}_i$). Let us denote by

$$W_1, \ldots, W_l$$

the irreducible cones that appear as an irreducible component of some of the k asymptotic cones.

Then one proves that the necessary and sufficient condition for Ω open and convex to have $H^1(\Omega, \mathcal{A}_0) = 0$ is that

« for any convex compact set $K \subset \Omega$ we can find another compact convex set K' with $K \subset K' \subset \Omega$ and a $\delta > 0$ such that for any j with $1 \leqslant j \leqslant l$ and for any plurisubharmonic function φ on $\mathbf{C}^n$ which verifies

$$\begin{cases} \varphi(\xi) \leqslant H_K(\xi) + \delta|\xi| & \forall \xi \in W_j \\ \varphi(\xi) \leqslant 0 & \forall \xi \in W_j \cap i\mathbf{R}^n \end{cases}$$

we also have

$$\varphi(\xi) \leqslant H_{K'}(\xi) \quad \forall \xi \in W_j \text{ ».}$$

This generalizes the theorem of Hörmander. The two essential tools that enable this generalization are the recourse to the cohomological formulation of the problem as a vanishing theorem of cohomology and the proper generalization of the characteristic variety of the overdetermined system $A_0(D)u = f$ via the primary decomposition of the module N.

One establishes then that if an algebraic cone W_j « admits a Phragmén-Lindelöf principle » the real part at every real point $a \neq 0$ is pure dimensional of real dimension equal to the complex dimension.

In the case dim $W_j = 2$ one proves that at any real point $a \neq 0$ every irreducible germ of W_j is non singular. One establishes that for « locally hyperbolic cones » (cf. [1], see definition in section 18) the Phragmén-Lindelöf principle holds when $\Omega = \mathbb{R}^n$ and finally one is able to establish a generalization of De Giorgi-Cattabriga theorem (section 19) to the case that the characteristic varieties have all dimension not greater than one.

Except for the generalization of De Giorgi-Cattabriga theorem all these applications extend arguments given already by Hörmander for the case of one operator.

The generalization of Phragmén-Lindelöf principle to overdetermined systems was formulated in the paper of Tetsuji Miwa: *On the global existence of real analytic solutions of systems of linear differential equations with constant coefficients* (Proc. Japan Acad., **49** (1973), pp. 500-502).

CONTENTS

SECTION 1

Preliminaries

a) Let $(x_1, \ldots, x_n)$ be cartesian coordinates in $\mathbb{R}^n$ and let ω be open in $\mathbb{R}^n$. We consider in $\mathbb{R}^n$ a Hilbert complex

$$(1) \qquad (\mathcal{E}^*(\omega), A_*) \equiv \left\{ \mathcal{E}^{p_0}(\omega) \xrightarrow{A_0(D_x)} \mathcal{E}^{p_1}(\omega) \xrightarrow{A_1(D_x)} \mathcal{E}^{p_2}(\omega) \longrightarrow \ldots \right\}$$

where $D_x = (\partial/\partial x_1, \ldots, \partial/\partial x_n)$.

Let $N = n + k$ and let $(x_1, \ldots, x_n, y_1, \ldots, y_k)$ be cartesian coordinates in $\mathbb{R}^N$; let $\tilde{\Omega}$ be open in $\mathbb{R}^N$. We consider in $\mathbb{R}^N$ an elliptic and Cauchy-Kowalewska suspension of the complex (1)

$$(2) \qquad (\mathcal{E}^*(\Omega), S_*) \equiv \left\{ \mathcal{E}^{s_0}(\Omega) \xrightarrow{S_0(D_x, D_y)} \mathcal{E}^{s_1}(\Omega) \xrightarrow{S_1(D_x, D_y)} \mathcal{E}^{s_2}(\Omega) \longrightarrow \ldots \right\}$$

where $D_y = (\partial/\partial y_1, \ldots, \partial/\partial y_k)$. We identify $\mathbb{R}^n$ with the subspace of $\mathbb{R}^N$: $\mathbb{R}^n = \{(x, y) \in \mathbb{R}^N | y = 0\}$. The operator $S_0(D_x, D_y) = S_0(D)$ will be an elliptic operator with constant coefficients and of Cauchy-Kowalewska type with respect to $\mathbb{R}^n$. The complex (2) will be a Hilbert complex in $\mathbb{R}^N$ and if we set $\mathcal{P}_N = \mathbb{C}[\xi_1, \ldots, \xi_n, \eta_1, \ldots, \eta_k]$, $\mathcal{P}_n = \mathbb{C}[\xi_1, \ldots, \xi_n]$ the above situation will arise from a commutative diagram of $\mathcal{P}_n$-homomorphisms with exact rows.

$$\begin{array}{ccccccccc}
0 \leftarrow M \leftarrow & \mathcal{P}_N^{s_0} & \xrightarrow{{}^tS_0(\xi, \eta)} & \mathcal{P}_N^{s_1} & \xrightarrow{{}^tS_1(\xi, \eta)} & \mathcal{P}_N^{s_2} & \leftarrow \ldots \\
\Vert \qquad & \downarrow {}^t\tau_0 & & \downarrow {}^t\tau_1 & & \downarrow {}^t\tau_2 & \\
0 \leftarrow (M)_n \leftarrow & \mathcal{P}_n^{p_0} & \xrightarrow{{}^tA_0(\xi)} & \mathcal{P}_n^{p_1} & \xrightarrow{{}^tA_1(\xi)} & \mathcal{P}_n^{p_2} & \rightarrow \ldots
\end{array}$$

The first row is also an exact sequence of $\mathcal{P}_N$-homomorphisms, M is a $\mathcal{P}_N$-module of finite type and $(M)_n$ denotes the module M considered as a $\mathcal{P}_n$-module via the natural injection $\mathcal{P}_n \rightarrow \mathcal{P}_N$.

The maps ${}^t\tau_j$ are given by matrices with entries in $\mathcal{P}_N$ and for $\omega = \Omega \cap \mathbb{R}^n$

define linear maps

$$\tau_j(D_y, D_x)\colon \mathcal{E}^{s_j}(\Omega) \to \mathcal{E}^{p_j}(\omega) \qquad j = 0, 1, \ldots$$

by

$$f \to \big(\tau_j(D_x, D_y) f\big)\big|_{y=0} .$$

We denote by $\mathcal{E}_{S_0}$ the sheaf of germs of C^∞ functions in $\mathbb{R}^N$, $f \in \mathcal{E}^{s_0}$ (with values in $\mathbb{C}^{s_0}$), which satisfy the equation $S_0(D) f = 0$. Similarly replacing the sheaf $\mathcal{E}$ of C^∞ functions by the sheaf $\mathcal{A}$ of real analytic functions (complex valued) we define the sheaf $\mathcal{A}_{S_0}$ of germs of real analytic functions (with values in $\mathbb{C}^{s_0}$) $f \in \mathcal{A}^{s_0}$ such that $S_0(D) f = 0$.

Finally we consider in $\mathbb{R}^n$ the sheaf $\mathcal{A}_{A_0}$ of real analytic functions $f \in \mathcal{A}^{p_0}$ which satisfy the equation $A_0(D) f = 0$ (where here $\mathcal{A}$ denotes the sheaf of complex valued real analytic functions in $\mathbb{R}^n$).

Because the operator $S_0(D)$ is elliptic we have an isomorphism

$$\mathcal{E}_{S_0} \simeq \mathcal{A}_{S_0} .$$

Also because $S_0(D)$ is of Cauchy-Kowalewska type with respect to $\mathbb{R}^n$ we have an isomorphism

$$\tau_0(D_x, D_y)\colon \mathcal{A}_{S_0} \xrightarrow{\sim} \mathcal{A}_{A_0}$$

so that combining the two isomorphisms we get an isomorphism

$$« \tau_0 »\colon \mathcal{E}_{S_0} \xrightarrow{\sim} \mathcal{A}_{A_0} .$$

b) Let $H = n + h$ and let $(x_1, \ldots, x_n, t_1, \ldots, t_h)$ be cartesian coordinates in $\mathbb{R}^H$. Let G be open in $\mathbb{R}^H$. We consider in $\mathbb{R}^H$ another Cauchy-Kowalewska elliptic suspension of the complex (1)

$$(3) \qquad (\mathcal{E}^*(G), L_*) \equiv \Big\{\{\mathcal{E}^{\varrho_0}(G) \xrightarrow{L_0(D_y, D_t)} \mathcal{E}^{\varrho_1}(G) \xrightarrow{L_1(D_y, D_t)} \mathcal{E}^{\varrho_2}(G) \to \ldots\Big\}$$

where $D_t = (\partial/\partial t_1, \ldots, \partial/\partial t_h)$ and where $\mathbb{R}^n$ is identified with the subspace of $\mathbb{R}^H$ $\{(x, t) \in \mathbb{R}^H | t = 0\}$. With obvious analogous notations to the ones used above we have a natural isomorphism of sheaves

$$« \lambda_0 »\colon \mathcal{E}_{L_0} \xrightarrow{\sim} \mathcal{A}_{A_0}$$

where $\mathcal{E}_{L_0}$ is the sheaf of germs of functions $f \in \mathcal{E}^{\varrho_0}$ such that $L_0(D) f = 0$.

LEMMA 1. *Let ω be open in $\mathbb{R}^n$ and let Ω be an open neighborhood of ω in $\mathbb{R}^N$. There exists an ω-connected open neighborhood G of ω in $\mathbb{R}^H$ such that*

for every $u \in \Gamma(\Omega, \mathcal{E}_{S_0})$ there exists a unique $v \in \Gamma(G, \mathcal{E}_{L_0})$ with

$$\tau_0(u) = \lambda_0(v) .$$

PROOF. We consider $\mathbb{R}^n \subset \mathbb{R}^N$ and $\mathbb{R}^n \subset \mathbb{R}^H$ imbedded in a natural way into $\mathbb{C}^n \subset \mathbb{C}^N$ and $\mathbb{C}^n \subset \mathbb{C}^H$ respectively as their real parts. From lemma 1 of section 2 of [4] we deduce that Ω has an open Ω-connected neighborhood $\tilde{\Omega}$ in $\mathbb{C}^N$ such that every $u \in \Gamma(\Omega, \mathcal{E}_{S_0})$ extends to a holomorphic function $\tilde{u}$ on $\tilde{\Omega}$.

Let $\tilde{\omega} = \tilde{\Omega} \cap \mathbb{C}^n$. This is an open neighborhood of ω in $\mathbb{C}^n$ and for every $u \in \Gamma(\Omega, \mathcal{E}_{S_0})$ $\tau_0(u)$ extends to a holomorphic function in $\tilde{\omega}$ because of the way τ_0 is defined (as the restriction to ω of a differential operator with constant coefficients applied to u).

By virtue of lemma 2 of [4] there exists an open neighborhood $\tilde{G}$ of ω in $\mathbb{C}^H$ with the following property: the Cauchy problem

$$\begin{cases} L_0(D)v = 0 & v \in \mathcal{A}^{q_0}(G) \\ \lambda_0(v) = w & \\ A_0(D)w = 0 & w \in \mathcal{A}^{p_0}(\omega) \end{cases}$$

for $G = \tilde{G} \cap \mathbb{R}^H$ when w is defined and holomorphic in $\tilde{\omega}$ has a unique solution v which extends holomorphically to $\tilde{G}$.

In particular we can take $w = \tau_0(u)$ so that given $u \in \Gamma(\Omega, \mathcal{E}_{S_0})$ there exists a unique $v \in \Gamma(G, \mathcal{E}_{L_0})$ with $\tau_0(u) = \lambda_0(v)$.

REMARK. Consider in $\mathbb{C}^{m+1} = \mathbb{C}^m \times \mathbb{C}$, where $(z_1, .., z_m, t)$ are holomorphic coordinates, the closed strip, for $0 < R < 1$,

$$\overline{P} = \{(z, t) \in \mathbb{C}^{m+1} \big| \, \|z\| \leqslant R\}$$

where $\|z\| = \sup_{1 \leqslant i \leqslant m} |z_i|$. We set $\overline{P}_0 = \{(z, t) \in \overline{P} | t = 0\}$. Consider in $\overline{P}$ a Cauchy-Kowalewska equation with constant coefficients in one unknown function u

$$(\alpha) \qquad D_t^l u = \sum_{\substack{\beta + |\alpha| \leqslant l \\ \beta < l}} c_{\alpha\beta} D_z^\alpha D_t^\beta u$$

with the initial conditions

$$(\alpha_0) \qquad u(z, 0) = \nu_0(z) , \qquad \frac{\partial u}{\partial t}(z, 0) = \nu_1(z), \ldots, \frac{\partial^{l-1} u}{\partial t}(z, 0) = \nu_{l-1}(z)$$

when the ν_j's are holomorphic in a neighborhood of $\overline{P}_0$.

Then this Cauchy problem admits a unique holomorphic solution u in a part $\Omega \subset \overline{P}$ defined by an inequality of the form

$$|t| < c(R - \|z\|)$$

where $c > 0$ is a constant depending only on the coefficients $c_{\alpha\beta}$. Moreover given any compact subset $K \subset \Omega$ there exists a constant $c_K > 0$ such that for the solution we have the estimate in terms of the Cauchy data

$$\|u\|_K \leqslant c_K \|\nu\|_{\overline{P}_0}$$

where $\|u\|_K = \sup_K |u|$ and $\|\nu\|_{\overline{P}_0} = \sup_{0 \leqslant i \leqslant l-1} \sup_{\overline{P}_0} |\nu_i|$.

This follows for instance from the proof of the Cauchy-Kowalewska theorem as given in [3] § 3.

We now apply this remark to the previous argument. We deduce then that the map

$$\alpha\colon \Gamma(\Omega, \mathcal{E}_{S_0}) \to \Gamma(G, \mathcal{E}_{L_0})$$

which associates to $u \in \Gamma(\Omega, \mathcal{E}_{S_0})$ the element $\alpha(u) = \nu \in \Gamma(G, \mathcal{E}_{L_0})$ for which we have $\tau_0(u) = \lambda_0(\nu)$ on ω, is a *continuous map*, provided G is chosen sufficiently small.

Indeed lemma 2 of [4] is based on a Cauchy-Kowalewska argument (derived from the sufficiency part of theorem 2 of [4]) of the type mentioned above.

c) We consider $\mathbb{R}^n$ imbedded as its real part in $\mathbb{C}^n$ and on $\mathbb{C}^n$ the sheaf $\mathcal{O}_{A_0}$ of germs of holomorphic functions u, with values in $\mathbb{C}^{p_0}$, which satisfy the equation $A_0(D_z)u = 0$ where $D_z = (\partial/\partial z_1, \dots, \partial/\partial z_n)$, $z_j = x_j + is_j$, $1 \leqslant j \leqslant n$, being complex coordinates in $\mathbb{C}^n$.

Similarly we consider $\mathbb{R}^N$ imbedded in $\mathbb{C}^N$ and on $\mathbb{C}^N$ we consider the sheaf $\mathcal{O}_{S_0}$ of germs of holomorphic functions u, with values in $\mathbb{C}^{s_0}$ which satisfy the equation $S_0(D_z, D_\xi)u = 0$ where $D_\xi = (\partial/\partial\xi_1, \dots, \partial/\partial\xi_k)$, $\xi_j = y_j + it_j$, $1 \leqslant j \leqslant k$, being a system of complex coordinates that with the coordinates z_j $(1 \leqslant j \leqslant n)$ complete the complex coordinates from $\mathbb{C}^n$ to $\mathbb{C}^N$.

LEMMA 2. *The natural restriction map*

$$r\colon \Gamma(\mathbb{C}^N, \mathcal{O}_{S_0}) \to \Gamma(\mathbb{R}^N, \mathcal{E}_{S_0})$$

is an isomorphism.

PROOF. We have to show that every $u \in \Gamma(\mathbb{R}^N, \mathcal{E}_{S_0})$ is the trace on $\mathbb{R}^N$ of an entire solution in $\mathbb{C}^N$ of the equation $S_0 u = 0$.

Since S_0 is elliptic there exists an open neighborhood U of $\mathbb{R}^N$ in $\mathbb{C}^N$ such that the natural restriction $\Gamma(U, \mathcal{O}_{S_0}) \to \Gamma(\mathbb{R}^N, \mathcal{E}_{S_0})$ is an isomorphism (cf. [4] corollary 2 to theorem 2).

Since S_0 has constant coefficients it is translation invariant, therefore it is not restrictive to assume that U is of the form

$$U(\varepsilon) = \left\{(z, \xi) \in \mathbb{C}^N \Big| \sum_1^n s_j^2 + \sum_1^k t_j^2 < \varepsilon\right\}$$

for some $\varepsilon = \varepsilon_0 > 0$.

Set

$$\Lambda = \sup\left\{\varepsilon > 0 \,|\, \Gamma(U(\varepsilon), \mathcal{O}_{S_0}) \xrightarrow{\sim} \Gamma(\mathbb{R}^N, \mathcal{E}_{S_0})\right\}.$$

We claim that $\Lambda = +\infty$ and this will prove the lemma. Suppose if possible that $\varepsilon_0 \leqslant \Lambda < +\infty$. Let $z_0 \in U(\Lambda - \varepsilon_0/2)$ and let $u \in \Gamma(U(\Lambda), \mathcal{O}_{S_0})$. Then $u|z_0 + \mathbb{R}^N$ extends to a holomorphic function $\tilde{u}$ in $z_0 + U(\varepsilon_0)$ which is a solution of $S_0 \tilde{u} = 0$. Now $\tilde{u} = u$ on $W = U(\Lambda) \cap (z_0 + U(\varepsilon_0))$ as two holomorphic functions defined in a connected open neighborhood W of $z_0 + \mathbb{R}^N$ and which agree on $z_0 + \mathbb{R}^N$, agree on W. Therefore u extends holomorphically to $U(\Lambda + \varepsilon_0/2)$. This contradicts the assumption $\Lambda < +\infty$.

REMARK. Since S_0 is elliptic the space $\Gamma(\mathbb{R}^N, \mathcal{E}_{S_0})$ with the topology of uniform convergence on compact subsets of $\mathbb{R}^N$ is a complete space, thus a Fréchet space. Similarly $\Gamma(\mathbb{C}^N, \mathcal{O}_{S_0})$ with the topology of uniform convergence on compact subsets of $\mathbb{C}^N$ is a Fréchet space. As the map r is continuous for the considered topologies it is a topological isomorphism.

We consider now the linear map

$$\tau_0 \colon \Gamma(\mathbb{C}^N, \mathcal{O}_{S_0}) \to \Gamma(\mathbb{C}^n, \mathcal{O}_{A_0})$$

defined by

$$u \to \left(\tau(D_z, D_\xi) u\right)_{\xi=0}.$$

LEMMA 3. *The map*

$$\tau_0 \colon \Gamma(\mathbb{C}^N, \mathcal{O}_{S_0}) \to \Gamma(\mathbb{C}^n, \mathcal{O}_{A_0})$$

is an isomorphism.

PROOF. The map τ_0 is defined by a differential operator with constant coefficients $\tau_0(D_z, D_\xi)$ followed by the restriction to $\mathbb{C}^n$.

From the proof of theorem 2 of [4] (sufficiency part) it follows that there exists an open neighborhood U of $\mathbb{C}^n$ in $\mathbb{C}^N$ with the property that for any $u \in \Gamma(\mathbb{C}^n, \mathcal{O}_{A_0})$ we can find $v \in \Gamma(U, \mathcal{O}_{S_0})$ such that $\tau_0(v) = u$.

As τ_0, S_0, A_0 are invariant by parallel translation to $\mathbb{C}^n$ we may assume that

$$U = U(\varepsilon) = \left\{(z, \xi) \in \mathbb{C}^N \,\middle|\, \sum_{j=1}^{k} |\xi_j|^2 < \varepsilon\right\}$$

for some $\varepsilon = \varepsilon_0 > 0$.

Set

$$\Lambda = \sup \{\varepsilon > 0 \,|\, \tau_0 \colon \Gamma(U(\varepsilon), \mathcal{O}_{S_0}) \xrightarrow{\sim} \Gamma(\mathbb{C}^n, \mathcal{O}_{A_0})\} \,.$$

We claim that $\Lambda = +\infty$ and this will prove the lemma.

Suppose if possible $\varepsilon_0 \leqslant \Lambda < +\infty$, and let $z_0 \in U(\Lambda - \varepsilon_0/2)$. With obvious notations we will have

$$\tau_0(D_z, D_\xi)\, v|_{z_0 + \mathbb{C}^n} \in \Gamma(z_0 + \mathbb{C}^n, \mathcal{O}_{A_0})$$

for every $v \in \Gamma(U(\Lambda), \mathcal{O}_{S_0})$. Therefore v extends to a holomorphic function $\tilde{v}$ defined on $z_0 + U(\varepsilon_0)$ which is a solution of $S_0 \tilde{v} = 0$. But now two holomorphic functions v_1 and v_2 defined in an open connected neighborhood W of $z_0 + \mathbb{C}^n$ in $\mathbb{C}^N$, solutions of the equation $S_0 v = 0$ and such that

$$(\tau_0(D_z, D_\xi) v_1)|z_0 + \mathbb{C}^n = (\tau_0(D_z, D_\xi) v_2)|z_0 + \mathbb{C}^n$$

must coincide on W. It follows that every $v \in \Gamma(U(\Lambda), O_{S_0})$ extends holomorphically to $U(\Lambda + \varepsilon_0/2)$. This contradicts the assumption $\Lambda < +\infty$.

REMARK. With the topology of uniform convergence on compact sets the spaces $\Gamma(\mathbb{C}^N, \mathcal{O}_{S_0})$ and $\Gamma(\mathbb{C}^n, \mathcal{O}_{A_0})$ are Fréchet spaces. The map τ_0 of the previous lemma is continuous and thus it is a topological isomorphism.

PROPOSITION 0. *Let* (2) *be an elliptic Cauchy-Kowalewska suspension of the complex* (1).

Let ω be open and convex in $\mathbb{R}^n$ and let Ω be an open neighborhood of ω in $\mathbb{R}^N$.

There is an open neighborhood Ω' of ω in $\mathbb{R}^N$ with $\Omega' \subset \Omega$ and with the property:

for every $f \in \Gamma(\Omega, \mathcal{E}_{S_0})$ one can find a sequence $\{f_\nu\}_{\nu \in \mathbb{N}} \subset \Gamma(\mathbb{R}^N, \mathcal{E}_{S_0})$ such that $\{f_\nu\}$ converges to f uniformly on any compact subset $K \subset \Omega'$.

Proof. Let (3) be another elliptic and Cauchy-Kowalewska suspension of the complex (1) to $\mathbb{R}^{n+1}$ ($H = n + 1$, $h = 1$). Let $x_1, \ldots, x_n$, t denote cartesian coordinates in $\mathbb{R}^{n+1}$ as stipulated before. By lemma 1 we can find an open neighborhood G of ω in $\mathbb{R}^{n+1}$ and an open neighborhood Ω' of ω in $\mathbb{R}^N$ and continuous maps

$$\Gamma(\Omega, \mathcal{E}_{S_0}) \xrightarrow{\alpha} \Gamma(G, \mathcal{E}_{L_0}) \xrightarrow{\beta} \Gamma(\Omega', \mathcal{E}_{S_0})$$

where α is defined by

$$\tau_0(u) = \lambda_0(\alpha(u)) \qquad \forall u \in \Gamma(\Omega, \mathcal{E}_{S_0})$$

and β is defined by

$$\lambda_0(v) = \tau_0(\beta(v)) \qquad \forall v \in \Gamma(G, \mathcal{E}_{L_0}) .$$

We may assume that Ω' is connected and that $\Omega' \subset \Omega$ so that the composition map $\beta \circ \alpha$ must coincide with the natural restriction map from Ω to Ω'. Also we may assume that G is an open set of the form

$$G = \{(x, t) \in \mathbb{R}^{n+1} | x \in \omega, \; |t| < \varrho(x)\}$$

for some continuous function $\varrho(x) > 0$ defined on ω.

Given $f \in \Gamma(\Omega, \mathcal{E}_{S_0})$ set $g = \alpha(f)$. By theorem 5 of [4] we can find a sequence $\{g_\nu\}_{\nu \in \mathbb{N}} \in \Gamma(\mathbb{R}^{n+1}, \mathcal{E}_{L_0})$ such that $\{g_\nu\}$ converges uniformly to g on any compact subset of G.

Set $f_\nu = \beta(g_\nu)$. By lemmas 2 and 3 we have $f_\nu \in \Gamma(\mathbb{R}^N, \mathcal{E}_{S_0})$. Also we have $f = \beta(g)$ on Ω'. Since β is continuous, on every compact subset $K \subset \Omega'$, $\{f_\nu\}$ must converge uniformly to f.

Corollary. *Let $\omega_1 \subset \omega_2$ be two open convex sets in $\mathbb{R}^n$. Let Ω_i, $i = 1, 2$, be an open neighborhood of ω_i in $\mathbb{R}^N$. There exist open neighborhoods $\Omega_i' \subset \Omega_i$ of ω_i in $\mathbb{R}^N$ depending only on Ω_i, $i = 1, 2$, such that*

for every $f \in \Gamma(\Omega_1 \cup \Omega_2, \mathcal{E}_{S_0})$ we can find a sequence $\{f_\nu\} \subset \Gamma(\mathbb{R}^N, \mathcal{E}_{S_0})$ which converges uniformly to f on any compact set $K \subset \Omega_1' \cup \Omega_2'$.

Proof. We determine from Ω_i, G_i and Ω_i' as in the previous proposition. We may assume that

$$G_i = \{(x, t) \in \mathbb{R}^{m+1} | x \in \omega_i, \; |t| < \varrho_i(x)\}$$

where $\varrho_i(x) > 0$ is defined and lower semicontinuous on ω_i. Let $G = G_1 \cup G_2$ then G is defined by

$$G = \{(x, t) \in \mathbb{R}^{n+1} | x \in \omega_2, |t| < \varrho(x)\}$$

where $\varrho(x) = \varrho_2(x)$ on $\omega_2 - \omega_1$ and $\varrho(x) = \sup(\varrho_1(x), \varrho_2(x))$ for $x \in \omega_1$. Then $\varrho(x)$ is lower semicontinuous and $\varrho(x) > 0$ on ω_2. Set $g = \alpha(f)$ where α is defined as in the previous proposition. Then g is defined in G and by theorem 5 of [4] we can find a sequence $\{g_\nu\} \subset \Gamma(\mathbb{R}^{n+1}, \mathcal{E}_{L_0})$ which converges uniformly to g on any compact subset of G.

If β is defined as in the previous proposition, we have $\beta(g) = f$ on $\Omega_1' \cup \Omega_2'$ and $f_\nu = \beta(g_\nu) \in \Gamma(\mathbb{R}^N, \mathcal{E}_{S_0})$. Moreover by the continuity of β we derive that the sequence $\{f_\nu\}$ converges uniformly to f on any compact set $K \subset \Omega_1' \cup \Omega_2'$.

d) For ω open in $\mathbb{R}^n$ and for $T > 0$ we set

$$C_N(\omega, T) = \{(x, y) \in \mathbb{R}^N | x \in \omega, |y| < T\}$$

where $|y| = \left(\sum_1^k y_j^2\right)^{\frac{1}{2}}$;

$$C_H(\omega, T) = \{(x, t) \in R^H | x \in \omega, |t| < T\}$$

where $|t| = \left(\sum_1^h t_j^2\right)^{\frac{1}{2}}$. If ω is convex also $C_N(\omega, T)$ and $C_H(\omega, T)$ are convex. If we are dealing with only the suspension (2) of (1) to $\mathbb{R}^N$ the sets $C_N(\omega, T)$ will also be denoted by $C(\omega, T)$.

SECTION 2

The Splitting Condition

Let a Hilbert complex (1) be given in $\mathbb{R}^n$ with an elliptic Cauchy-Kowalewska supension (2) to $\mathbb{R}^N$.

Let ω be an open (non empty) *convex* set in $\mathbb{R}^n$. We will say that *on ω the splitting condition is satisfied* with respect to the elliptic Cauchy-Kowalewska suspension (2) if the following holds

(S) *For every non empty convex open subset $\omega_1 \subset\subset \omega$ we can find a real $\delta > 0$ and an open convex set ω_2 with $\omega_1 \subset \omega_2 \subset\subset \omega$ with the property: for every real T with $0 < T < \delta$ we can find an open neighborhood $\Omega(T) \subset C(\omega, T)$ of ω in $\mathbb{R}^N$ such that for any $u \in \Gamma(C(\omega_2, T), \mathcal{E}_{S_0})$ there exist $u_1 \in \Gamma(C(\omega_1, \delta), \mathcal{E}_{S_0})$ and $u_2 \in \Gamma(\Omega(T), \mathcal{E}_{S_0})$ with $u = u_1 - u_2$ on $\Omega(T) \cap C(\omega_1, T)$.*

Proposition 1. *The splitting conditions (S) for a non empty open convex set ω in $\mathbb{R}^n$ is equivalent to the following condition:*

For every non empty open convex set $\omega_1 \subset\subset \omega$ we can find a real $\delta > 0$ and an open convex set ω_2 with $\omega_1 \subset \omega_2 \subset\subset \omega$ with the property

for every real T with $0 < T < \delta$ we can find an open neighborhood $\Omega(T) \subset C(\omega, T)$ of ω in $\mathbb{R}^N$ such that the natural restriction map

$$H^1(C(\omega, T) \cup C(\omega_2, \delta), \mathcal{E}_{S_0}) \to H^1(\Omega(T) \cup C(\omega_1, \delta), \mathcal{E}_{S_0})$$

has zero image.

Proof. We consider $\{C(\omega, T), C(\omega_2, \delta)\}$ and $\{\Omega(T), C(\omega_1, \delta)\}$ as open coverings of $C(\omega, T) \cup C(\omega_2, \delta)$ and $\Omega(T) \cup C(\omega_1, \delta)$ respectively. We have $\Omega(T) \subset C(\omega, T)$, $C(\omega_1, \delta) \subset C(\omega_2, \delta)$ and these inclusions define a natural map λ of the alternate cocycle groups on those coverings (denoted by Z^1)

$$Z^1(\{C(\omega, T), C(\omega_2, \delta)\}, \mathcal{E}_{S_0}) \xrightarrow{\lambda} Z^1(\{\Omega(T), C(\omega_1, \delta)\}, \mathcal{E}_{S_0}) \,.$$

We have a natural commutative diagram

$$\begin{array}{ccc} Z^1(\{C(\omega, T), C(\omega_2, \delta)\}, \mathcal{E}_{S_0}) & \xrightarrow{\lambda} & Z^1(\{\Omega(T), C(\omega_1, \delta)\}, \mathcal{E}_{S_0}) \\ \downarrow \alpha & & \downarrow \beta \\ H^1(C(\omega, T) \cup C(\omega_2, \delta), \mathcal{E}_{S_0}) & \xrightarrow{\mu} & H^1(\Omega(T) \cup C(\omega_1, \delta), \mathcal{E}_{S_0}) \end{array}$$

where α and β induce injective maps on cohomology by a general property of the first Čech cohomology groups. Now $C(\omega, T)$ and $C(\omega_2, \delta)$ are open convex sets. Therefore $\{C(\omega, T), C(\omega_2, \delta)\}$ is a Leray covering of $C(\omega, T) \cup \cup C(\omega_2, \delta)$ and consequently α is also a surjective map. Every cohomology class of $H^1(C(\omega, T) \cup C(\omega_2, \delta), \mathcal{E}_{S_0})$ is therefore represented by an element

$$u \in Z^1(\{C(\omega, T), C(\omega_2, \delta)\}, \mathcal{E}_{S_0}) \simeq \Gamma(C(\omega_2, T), \mathcal{E}_{S_0}) .$$

To say that $\mu(\{u\}) = 0$ is equivalent to say that $\beta \circ \lambda(u) = 0$ and this means the existence of $u_1 \in \Gamma(C(\omega_1, \delta), \mathcal{E}_{S_0})$, $u_2 \in \Gamma(\Omega(T), \mathcal{E}_{S_0})$ such that $u = u_1 - u_2$ on $\Omega(T) \cap C(\omega_1, \delta) \cap C(\omega_2, T) = \Omega(T) \cap C(\omega_1, T)$. This proves our contention.

Remark. Note that for every $\delta' > \delta$ we do have

$$C(\omega, T) \cap C(\omega_2, \delta') = C(\omega, T) \cap C(\omega_2, \delta) .$$

Therefore in the conclusion of the previous condition the space $H^1(C(\omega, T) \cup \cup C(\omega_2, \delta), \mathcal{E}_{S_0})$ can be replaced by $H^1(C(\omega, T) \cup C(\omega_2, \delta'), \mathcal{E}_{S_0})$ for any choice of $\delta' \geqslant \delta$.

Proposition 2. *Let ω be an open convex set in $\mathbb{R}^n$. If ω satisfies the splitting condition* (S) *with respect to an elliptic Cauchy-Kowalewska suspension* (2) *of the complex* (1), *then ω satisfies the splitting condition* (S) *with respect to any other elliptic Cauchy-Kowalewska suspension* (3) *of the complex* (1).

Proof. Let $\omega_1 \subset\subset \omega$ be an open non empty and convex set in ω. Let us select an open convex set ω_1' with

$$\omega_1 \subset\subset \omega_1' \subset\subset \omega .$$

By the assumption there exist $\delta' > 0$ and an open convex set ω_2' with $\omega_1' \subset \omega_2' \subset\subset \omega$ with the property that for every T' with $0 < T' < \delta'$ one can find an open neighborhood $\Omega'(T') \subset C_N(\omega, T')$ of ω in $\mathbb{R}^N$ such that

for any $u \in \Gamma(C_N(\omega_2', T'), \mathcal{E}_{S_0})$ there exist

$$u_1 \in \Gamma(C_N(\omega_1', \delta'), \mathcal{E}_{S_0}) \quad \text{and} \quad u_2 \in \Gamma(\Omega'(T'), \mathcal{E}_{S_0})$$

with

$$u = u_1 - u_2 \quad \text{on } \Omega'(T') \cap C_N(\omega_1', T') .$$

By virtue of lemma 1 there exist an open neighborhood V of ω_1' in $\mathbb{R}^H$ such that for every $u \in \Gamma(C_N(\omega_1', \delta'), \mathcal{E}_{S_0})$ we can find an unique $v \in \Gamma(V, \mathcal{E}_{L_0})$ with

$$\tau_0(u) = \lambda_0(v) \quad \text{on } \omega_1' .$$

We choose now $\delta > 0$ such that

$$C_H(\omega_1, \delta) \subset V .$$

This is possible as $\omega_1 \subset\subset \omega_1'$. We also choose ω_2 convex open in $\mathbb{R}^N$ such that

$$\omega_2' \subset\subset \omega_2 \subset\subset \omega .$$

Let T be such that $0 < T < \delta$. We can find a neighborhood U of ω_2 in $\mathbb{R}^N$ such that for any $v \in \Gamma(C_H(\omega_2, T), \mathcal{E}_{L_0})$ there exists an unique $u \in \Gamma(U, \mathcal{E}_{S_0})$ with

$$\tau_0(u) = \lambda_0(v) \quad \text{on } \omega_2 .$$

We can then choose $T' > 0$ with $T' < \delta'$ so small that

$$C_N(\omega_2', T') \subset U .$$

This because $\omega_2' \subset\subset \omega_2$. We then determine the neighborhood $\Omega'(T')$ as indicated at the beginning and we then can find a neighborhood $G(T)$ of ω in $\mathbb{R}^H$ such that for every $u \in \Gamma(\Omega'(T'), \mathcal{E}_{S_0})$ there exists a unique $v \in \Gamma(G(T), \mathcal{E}_{L_0})$ with

$$\tau_0(u) = \lambda_0(v) \quad \text{on } \omega .$$

Let now $v \in \Gamma(C_H(\omega_2, T), \mathcal{E}_{L_0})$ be given. We find $u \in \Gamma(U, \mathcal{E}_{S_0})$ with $\tau_0(u) = \lambda_0(v)$ on ω_2 as explained before. We consider u as defined in $C_N(\omega_2', T')$ and thus we can find $u_1 \in \Gamma(C_N(\omega_1', \delta'), \mathcal{E}_{S_0})$, $u_2 \in \Gamma(\Omega'(T'), \mathcal{E}_{S_0})$ with

$$u = u_1 - u_2 \quad \text{on } \Omega'(T') \cap C_N(\omega_1', T') .$$

We then construct $v_1 \in \Gamma(U, \mathcal{E}_{L_0})$ with $\tau_0(u_1) = \lambda_0(v_1)$ on ω_1' and similarly we construct $v_2 \in \Gamma(G(T), \mathcal{E}_{L_0})$ with $\tau_0(u_2) = \lambda_0(v_2)$ on ω.

We claim that we have

$$\nu = \nu_1 - \nu_2 \qquad \text{on } G(T) \cap C_H(\omega_1, T)$$

and ν_1 is defined on $C_H(\omega_1, \delta)$ as this is contained in V.

Indeed by restricting $G(T)$ we may assume, without loss of generality, that $G(T) \cap C_H(\omega_1, T)$ is connected. Now to show that $\nu = \nu_1 - \nu_2$ on the region considered it is enough to show that $\lambda_0(\nu) = \lambda_0(\nu_1) - \lambda_0(\nu_2)$ on ω_1 because ν, ν_1, ν_2 are real analytic functions and both ν and $\nu_1 - \nu_2$ solve the same Cauchy problem

$$\begin{cases} L_0(D)w = 0 & \text{on a neighborhood of } \omega_1 \text{ in } \mathbb{R}^H \\ \lambda_0(w) = \lambda_0(\nu) = \lambda_0(\nu_1) - \lambda_0(\nu_2) = \sigma & \\ A_0(D)\sigma = 0 & \text{on } \omega_1 . \end{cases}$$

Now we have by construction

$$\lambda_0(\nu) = \tau_0(u) = \tau_0(u_1) - \tau_0(u_2) = \lambda_0(\nu_1) - \lambda_0(\nu_2) .$$

This completes the proof.

REMARK. Actually in the previous argument we have proved the following statement:

Assume that for any non empty open convex set $\omega_1' \subset\subset \omega$ *we can find an open convex set* ω_2' *with* $\omega_1' \subset \omega_2' \subset\subset \omega$ *and a* $\delta' > 0$ *such that*

for any T' *with* $0 < T' < \delta'$ *we can find an open neighborhood* $\Omega(T') \subset \subset C_N(\omega, T')$ *of* ω *in* $\mathbb{R}^N$ *such that the natural restriction map*

$$H^1(C_N(\omega, T') \cup C_N(\omega_2', \delta'), \mathcal{E}_{S_0}) \to H^1(\Omega(T') \cup C_N(\omega_1', \delta'), \mathcal{E}_{S_0})$$

has zero image.

Then for any choice of a non empty open convex set $\omega_1 \subset\subset \omega_1'$ *and of an open convex set* ω_2 *with* $\omega_2' \subset\subset \omega_2 \subset\subset \omega$ *we can find* $\delta > 0$ *such that for any* T *with* $0 < T < \delta$ *we can construct an open neighborhood* $G(T) \subset C_H(\omega, T)$ *of* ω *in* $\mathbb{R}^H$ *such that the natural restriction map*

$$H^1(C_H(\omega, T) \cup C_H(\omega_2, \delta), \mathcal{E}_{L_0}) \to H^1(G(T) \cup C_H(\omega_1, \delta), \mathcal{E}_{L_0})$$

has zero image.

SECTION 3

Suspensions in $\mathbb{R}^{n+1}$

In the case of suspensions from $\mathbb{R}^n$ to $\mathbb{R}^{n+1}$ the splitting condition can be made more precise. We have the following

PROPOSITION 3. *We assume that the complex* (2) *is an elliptic and Cauchy-Kowalewska suspension of the complex* (1) *and that* $N = n + 1$. *Let* ω *be a non empty open convex set in* $\mathbb{R}^n$ *which satisfies the splitting condition* (S) *with respect to the given suspension* (2). *Then* ω *satisfies also the following more precise splitting condition*

(S′) *for every non empty open convex subset* $\omega_1 \subset\subset \omega$ *we can find a real* $\delta > 0$ *and an open convex subset* ω_2 *with*

$$\omega_1 \subset \omega_2 \subset\subset \omega$$

with the property:
for every real T *with* $0 < T < \delta$ *and for any*

$$u \in \Gamma(C(\omega_2, T), \mathcal{E}_{S_0})$$

there exist

$$u_1 \in \Gamma(C(\omega_1, \delta), \mathcal{E}_{S_0}) \quad \text{and } u_2 \in \Gamma(C(\omega, T), \mathcal{E}_{S_0})$$

with $u = u_1 - u_2$ *on* $C(\omega_1, T)$.

Before giving the proof of this proposition we remark that with the same argument given in proposition 1 we can state the following

PROPOSITION 4. *The splitting condition* (S′) *for a non empty open convex set* ω *in* $\mathbb{R}^n$ *is equivalent to the following condition. For every non empty convex set* $\omega_1 \subset\subset \omega$ *we can find a real* $\delta > 0$ *and an open convex set* ω_2 *with* $\omega_1 \subset \omega_2 \subset\subset \omega$ *with the property that for every real* T *with* $0 < T < \delta$ *the natural restriction map*

$$H^1(C(\omega, T) \cup C(\omega_2, \delta), \mathcal{E}_{S_0}) \to H^1(C(\omega, T) \cup C(\omega_1, \delta), \mathcal{E}_{S_0})$$

has zero image.

PROOF OF PROPOSITION 3. α) By assumption, given a non empty convex set $\omega_1 \subset\subset \omega$ we can find $\delta > 0$ and ω_2 open convex with $\omega_1 \subset \omega_2 \subset\subset \omega$ such that for every T with $0 < T < \delta$ there exists a neighborhood $\Omega(T) \subset C(\omega, T)$ of ω in $\mathbb{R}^{n+1}$ such that the natural restriction map

$$H^1(C(\omega, T) \cup C(\omega_2, \delta), \mathcal{E}_{S_0}) \to H^1(\Omega(T) \cup C(\omega_1, \delta), \mathcal{E}_{S_0})$$

has zero image.

We set

$$B(T) = \Omega(T) \cup C(\omega_1, \delta) .$$

Without loss of generality we may assume that there is a lower semicontinuous $\varrho: \omega \to \mathbb{R}$ with $\varrho > 0$ such that

$$B(T) = \{(x, y) \in \mathbb{R}^{n+1} | x \in \omega, |y| < \varrho(x)\} .$$

We define

$$B^+(T) = \{(x, y) \in \mathbb{R}^{n+1} | x \in \omega, y > -\varrho(x)\}$$

$$B^-(T) = \{(x, y) \in \mathbb{R}^{n+1} | x \in \omega, y < \varrho(x)\}$$

$$= \{(x, y) \in \mathbb{R}^{n+1} | (x, -y) \in B^+(T)\}$$

$$W^+(T) = \{(x, y) \in \mathbb{R}^{n+1} | x \in \omega, y > -T\} \cup \{(x, y) \in \mathbb{R}^{n+1} | x \in \omega_2, y > -T - \delta\}$$

$$W^-(T) = \{(x, y) \in \mathbb{R}^{n+1} | x \in \omega, y < T\} \cup \{(x, y) \in \mathbb{R}^{n+1} | x \in \omega_2, y < T + \delta\}$$

$$= \{(x, y) \in \mathbb{R}^{n+1} | (x, -y) \in W^+(T)\}$$

and we set

$$W(T) = W^+(T) \cap W^-(T) .$$

We note that $B^+(T) \cup B^-(T) = \omega \times \mathbb{R} = W^+(T) \cup W^-(T)$ is an open convex set. We consider the open set $\omega \times \mathbb{R}$ as union of the two open sets $B^+(T)$ and $B^-(T)$ or $W^+(T)$ and $W^-(T)$ and we write the corresponding Meyer-Vietoris cohomology sequences. These give the exact sequences

$$H^1(\omega \times \mathbb{R}, \mathcal{E}_{S_0}) \to H^1(B^+(T), \mathcal{E}_{S_0}) \oplus H^1(B^-(T), \mathcal{E}_{S_0}) \to$$

$$\to H^1(B(T), \mathcal{E}_{S_0}) \to H^2(\omega \times \mathbb{R}, \mathcal{E}_{S_0})$$

and

$$H^1(\omega\times\mathbb{R}, \mathcal{E}_{S_0}) \to H^1(W^+(T), \mathcal{E}_{S_0}) \oplus H^1(W^-(T), \mathcal{E}_{S_0}) \to$$
$$\to H^1(W(T), \mathcal{E}_{S_0}) \to H^2(\omega\times\mathbb{R}, \mathcal{E}_{S_0}) .$$

Because $\omega\times\mathbb{R}$ is convex $H^1(\omega\times\mathbb{R}, \mathcal{E}_{S_0}) = 0 = H^2(\omega\times\mathbb{R}, \mathcal{E}_{S_0})$. Therefore

$$H^1(B(T), \mathcal{E}_{S_0}) \simeq H^1(B^+(T), \mathcal{E}_{S_0}) \oplus H^1(B^-(T), \mathcal{E}_{S_0})$$
$$H^1(W(T), \mathcal{E}_{S_0}) \simeq H^1(W^+(T), \mathcal{E}_{S_0}) \oplus H^1(W^-(T), \mathcal{E}_{S_0}) .$$

By assumption the natural restriction map

$$H^1(W(T), \mathcal{E}_{S_0}) \to H^1(B(T), \mathcal{E}_{S_0})$$

has zero image. This because

$$\{(x, y) \in \mathbb{R}^{n+1} | x \in \omega, |y| < T\} \cap \{(x, y) \in \mathbb{R}^{n+1} | x \in \omega_2, |y| < T + \delta\} =$$
$$= C(\omega_2, T) = C(\omega, T) \cap C(\omega_2, \delta)$$

and because the two sets of which we consider the intersection are convex.

By virtue of the isomorphisms established above the natural restriction maps

$$H^1(W^+(T), \mathcal{E}_{S_0}) \to H^1(B^+(T), \mathcal{E}_{S_0})$$

and

$$H^1(W^-(T), \mathcal{E}_{S_0}) \to H^1(B^-(T), \mathcal{E}_{S_0})$$

have zero image.

β) We define for $\delta > 0$, $t \geqslant 0$,

$$L^-(\delta, t) = \{(x, y) \in \mathbb{R}^{n+1} | x \in \omega_1, y < \delta\} \cup \{(x, y) \in \mathbb{R}^{n+1} | x \in \omega, y < t\} .$$

For every T with $0 < T < \delta$ we have the inclusion

$$L^-(\delta, 0) \subset B^-(T) .$$

Therefore from the conclusion of point α) we deduce that the natural restriction map

$$H^1(W^-(T), \mathcal{E}_{S_0}) \to H^1(L^-(\delta, 0), \mathcal{E}_{S_0})$$

has zero image for any choice of T with $0 < T < \delta$.

Now the differential operator $S_0(D)$ has constant coefficients and therefore it is invariant by translation. From the above conclusion we deduce with a translation of length $t > 0$ in the direction of the y axis that the natural restriction map

$$H^1(W^-(T+t), \mathcal{E}_{S_0}) \to H^1(L^-(\delta, t), \mathcal{E}_{S_0})$$

has zero image, as for any $t > 0$ $L^-(\delta + t, t) \supset L^-(\delta, t)$.

Replacing $T + t$ by T and t by $T - \varepsilon$ with $0 < \varepsilon \leqslant T$, we can equivalently state that for any T with $0 < T < \delta$ and any ε with $0 < \varepsilon \leqslant T$ the natural restriction map

$$H^1(W^-(T), \mathcal{E}_{S_0}) \to H^1(L^-(\delta, T - \varepsilon), \mathcal{E}_{S_0})$$

has zero image.

From theorem 5 of [4] the natural restriction map

$$\Gamma(\mathbb{R}^{n+1}, \mathcal{E}_{S_0}) \to \Gamma(L^-(\delta, T - \varepsilon), \mathcal{E}_{S_0})$$

has a dense image. Therefore we can apply proposition 14 of [4] and conclude that the natural restriction map

$$H^1(W^-(T), \mathcal{E}_{S_0}) \to H^1(L^-(\delta, T), \mathcal{E}_{S_0})$$

has zero image.

γ) If we define $L^+(\delta, T) = \{(x, y) \in \mathbb{R}^{n+1} | (x, -y) \in L^-(\delta, T)\}$, the same type of argument given above proves that the natural restriction map

$$H^1(W^+(T), \mathcal{E}_{S_0}) \to H^1(L^+(\delta, T), \mathcal{E}_{S_0})$$

(for any T with $0 < T < \delta$) has zero image.

Set $L(\delta, T) = L^+(\delta, T) \cap L^-(\delta, T)$ and let us note that $L^+(\delta, T) \cup \cup L^-(\delta, T) = \omega \times \mathbb{R}$ is an open convex set. With the same argument given in point α) we deduce a natural isomorphism

$$H^1(L(\delta, T), \mathcal{E}_{S_0}) \simeq H^1(L^+(\delta, T), \mathcal{E}_{S_0}) \oplus H^1(L^-(\delta, T), \mathcal{E}_{S_0}) .$$

Since the restriction maps $H^1(W^\pm(T), \mathcal{E}_{S_0}) \to H^1(L^\pm(\delta, T), \mathcal{E}_{S_0})$ have zero image it follows that the restriction map

$$H^1(W(T), \mathcal{E}_{S_0}) \to H^1(L(\delta, T), \mathcal{E}_{S_0})$$

has also zero image (for $0 < T < \delta$).

Now

$$L(\delta, T) = C(\omega, T) \cup C(\omega_1, \delta)$$

and $W(T)$ is a union of two convex sets whose intersection is $C(\omega_2, T)$ as we have already remarked. From the last conclusion we deduce therefore that for any T, with $0 < T < \delta$, the natural restriction map

$$H^1(C(\omega, T) \cup C(\omega_2, \delta), \mathcal{E}_{S_0}) \to H^1(C(\omega, T) \cup C(\omega_1, \delta), \mathcal{E}_{S_0})$$

has zero image. Because of proposition 4 this is the statement we wanted to prove.

REMARK. The argument given for the proof of proposition 3 has actually established the following more precise statement

Let (2) *be an elliptic Cauchy-Kowalewska suspension of a Hilbert complex* (1) *and let* $N = n + 1$.

Let $\emptyset \neq \omega_1 \subset \omega_2 \subset\subset \omega$ *be open convex sets, let* $\delta > 0$, $T > 0$ *be given and let* $\Omega(T) \subset C(\omega, T)$ *be an open neighborhood of* ω *in* $\mathbb{R}^{n+1}$.

Assume that the natural restriction map

$$H^1(C(\omega, T) \cup C(\omega_2, \delta), \mathcal{E}_{S_0}) \to H^1(\Omega(T) \cup C(\omega_1, \delta), \mathcal{E}_{S_0})$$

has zero image.

Then also the natural restriction map

$$H^1(C(\omega, T) \cup C(\omega_2, \delta), \mathcal{E}_{S_0}) \to H^1(C(\omega, T) \cup C(\omega_1, \delta), \mathcal{E}_{S_0})$$

has zero image.

SECTION 4

Characterization of the Splitting Condition

Let (1) be a Hilbert complex in $\mathbb{R}^n$ of which we consider an elliptic Cauchy-Kowalewska suspension (2) to R^N. We have the following characterization of the splitting condition (S):

THEOREM 1. *Let ω be an open convex set in $\mathbb{R}^n$.*

The set ω satisfies the splitting condition (S) with respect to the suspension (2) of the complex (1) if and only if ω if analytically convex with respect to the complex (1) i.e. if and only if

$$H^1(\omega, \mathcal{A}_{A_0}) = 0 .$$

PROOF. (α) Let us assume that ω satisfies the splitting condition (S) and that ω is non empty.

By proposition 2 it is not restrictive to assume that $N = n + 1$, as there exists always an elliptic Cauchy-Kowalewska suspension of a given complex (1) with this property. We may therefore assume that ω satisfies the more precise splitting condition (S'). Let ω_1 be a non empty open convex relatively compact subset of ω. By successive application of the condition (S') we can construct

i) a sequence $\omega_1 \subset\subset \omega_2 \subset\subset \omega_3 \ldots \subset\subset \omega$ of open convex relatively compact subsets of ω with $\omega = \bigcup_{i=1}^{\infty} \omega_i$;

ii) a sequence $\delta_1 > \delta_2 > \delta_3 \ldots$ of positive numbers $\delta_i > 0$ for $i = 1, 2, 3, \ldots$ such that

for any choice of the index $i = 1, 2, 3, \ldots$ and any choice of T with $0 < T < \delta_i$ the natural restriction map

$$H^1(C(\omega, T) \cup C(\omega_{i+1}, \delta_i), \mathcal{E}_{S_0}) \to H^1(C(\omega, T) \cup C(\omega_i, \delta_i), \mathcal{E}_{S_0})$$

has zero image.

Let us select a sequence of positive numbers $\{T_i\}_{i=1,2,..}$ with

$$0 < T_i \leqslant \delta_i \qquad \text{for } i = 1, 2, \ldots \text{ and } T_1 \geqslant T_2 \geqslant T_3 \geqslant \ldots$$

and set

$$\Omega = \bigcup_{i=1}^{\infty} C(\omega_i, T_i)$$

$$\Omega' = \bigcup_{i=1}^{\infty} C(\omega_i, T_{i+1}) \, .$$

If we prove that the natural restriction map

$$H^1(\Omega, \mathcal{E}_{S_0}) \to H^1(\Omega', \mathcal{E}_{S_0})$$

has zero image we can conclude that $H^1(\omega, \mathcal{A}_{A_0}) = 0$. Indeed when the sequence $\{T_i\}$ varies Ω describes a fundamental system of neighborhoods of ω in $\mathbb{R}^{n+1}$ and we can apply proposition 10 of [4].

(β) Let ν be an integer $\geqslant 1$ and set

$$\Omega'_\nu = \bigcup_{i=1}^{\infty} C(\omega_i, T_{i+1}) \, .$$

By theorem 5 of [4] we have that the natural restriction map

$$\Gamma(\mathbb{R}^{n+1}, \mathcal{E}_{S_0}) \to \Gamma(\Omega'_\nu, \mathcal{E}_{S_0})$$

has a dense image. We can therefore, by application of proposition 14 of [4] conclude that in order to prove that the natural restriction map

$$H^1(\Omega, \mathcal{E}_{S_0}) \to H^1(\Omega', \mathcal{E}_{S_0})$$

has zero image, it is enough to show that for every $\nu \geqslant 1$ the natural restriction map

$$H^1(\Omega, \mathcal{E}_{S_0}) \to H^1(\Omega'_\nu, \mathcal{E}_{S_0})$$

has zero image.

Let $Z^1(\{C(\omega_i, T_i)\}_{i=1,2,\ldots}, \mathcal{E}_{S_0})$ denote the space of alternate one-cocycles on the covering $\{C(\omega_i, T_i)\}_{i=1,2,\ldots}$ with values in the sheaf $\mathcal{E}_{S_0}$. The sets $C(\omega_i, T_i)$ are open and convex and thus the considered covering is a Leray covering for the sheaf $\mathcal{E}_{S_0}$. Therefore every cohomology class $f \in H^1(\Omega, \mathcal{E}_{S_0})$

can be represented by a cocycle $\{f_{ij}\} \in Z^1(\{C(\omega_i, T_i)\}_{i=1,2,\ldots}, \mathcal{E}_{S_0})$. From the cocycle condition we deduce that for $i < j$

$$f_{ij} = \sum_{h=i}^{j-1} f_{h\,h+1} \qquad \text{on } C(\omega_i, T_j)$$

(cf. [4] section 6). The image of this cocycle in the cocycle group $Z^1(\{C(\omega_i, T_{i+1})\}_{1\leqslant i\leqslant \nu}, \mathcal{E}_{S_0})$ will be a coboundary if and only if the natural image of the cohomology class f in $H^1(\Omega'_\nu, \mathcal{E}_{S_0})$ is zero. This because $\{C(\omega_i, T_{i+1})\}_{1\leqslant i\leqslant \nu}$ is also a Leray covering of Ω'_ν with respect to the sheaf $\mathcal{E}_{S_0}$. Now by [4] lemma 3 we deduce that this will be the case if and only if we can find $u_h \in \Gamma(C(\omega_h, T_{h+1}), \mathcal{E}_{S_0})$ for $1 \leqslant h \leqslant \nu$ such that

$$f_{h\,h+1} = u_h - u_{h+1}$$

on $C(\omega_h, T_{h+1})$ for every h with $1 \leqslant h \leqslant \nu - 1$.

(γ) Note that $f_{h\,h+1}$ is defined on $C(\omega_h, T_{h+1})$. Therefore by the splitting condition given in (α) we can find, if $h > 1$,

$$u^{(h)}_{h-1} \in \Gamma(C(\omega_{h-1}, \delta_{h-1}), \mathcal{E}_{S_0}), \qquad v^{(h)}_{h-1} \in \Gamma(C(\omega, T_{h+1}), \mathcal{E}_{S_0})$$

such that

$$f_{h\,h+1} = u^{(h)}_{h-1} - v^{(h)}_{h-1} \qquad \text{on } C(\omega_{h-1}, T_{h+1}).$$

We remark that $u^{(h)}_{h-1} = f_{h\,h+1} + v^{(h)}_{h-1}$ by the right hand expression can be extended and defined on $C(\omega_h, T_{h+1})$ so that we get

$$u^{(h)}_{h-1} \in \Gamma(C(\omega_{h-1}, \delta_{h-1}) \cup C(\omega_h, T_{h+1}), \mathcal{E}_{S_0}).$$

Now if $h - 1 > 1$ as $u^{(h)}_{h-1}$ is defined on $C(\omega_{h-1}, \delta_{h-1})$ by the splitting condition we can find

$$u^{(h)}_{h-2} \in \Gamma(C(\omega_{h-2}, \delta_{h-2}), \mathcal{E}_{S_0}), \qquad v^{(h)}_{h-2} \in \Gamma(C(\omega, \delta_{h-1}), \mathcal{E}_{S_0})$$

such that

$$u^{(h)}_{h-1} = u^{(h)}_{h-2} - v^{(h)}_{h-2} \qquad \text{on } C(\omega_{h-2}, \delta_{h-1}).$$

We remark that $u^{(h)}_{h-2} = u^{(h)}_{h-1} + v^{(h)}_{h-2}$ by the right hand expression can be extended and defined on $C(\omega_{h-1}, \delta_{h-1}) \cup C(\omega_h, T_{h+1})$ so that we get

$$u^{(h)}_{h-2} \in \Gamma(C(\omega_{h-2}, \delta_{h-2}) \cup C(\omega_{h-1}, \delta_{h-1}) \cup C(\omega_h, T_{h+1}), \mathcal{E}_{S_0})$$

Arguing with $u^{(h)}_{h-2}$ as we did with $u^{(h)}_{h-1}$, if $h-2>1$, and so on we can construct two sequences

$$u^{(h)}_{h-2}, u^{(h)}_{h-3}, \dots, u^{(h)}_{1}$$

$$\nu^{(h)}_{h-2}, \nu^{(h)}_{h-3}, \dots, \nu^{(h)}_{1}$$

with

$$u_j^{(h)} \in \Gamma\big(C(\omega_j, \delta_j) \cup C(\omega_{j+1}, \delta_{j+1}) \cup \dots \cup C(\omega_{h-1}, \delta_{h-1}) \cup C(\omega_h, T_{h+1}), \mathcal{E}_{S_0}\big)$$

$$\nu_j^{(h)} \in \Gamma\big(C(\omega, \delta_{j+1}), \mathcal{E}_{S_0}\big)$$

and such that

$$u^{(h)}_{j+1} = u_j^{(h)} - \nu_j^{(h)} \qquad \text{on } C(\omega_j, \delta_{j+1})$$

for $1 \leqslant j \leqslant h-2$. We have therefore for $h \geqslant 2$

$$\begin{aligned} f_{h\,h+1} &= u^{(h)}_{h-1} - \nu^{(h)}_{h-1} \\ &= u^{(h)}_{h-2} - \nu^{(h)}_{h-1} - \nu^{(h)}_{h-2} = \dots \\ &= u_1^{(h)} - (\nu^{(h)}_{h-1} + \nu^{(h)}_{h-2} + \dots + \nu_1^{(h)}) \,. \end{aligned}$$

Set $u^{(h)} = u_1^{(h)}$, $\nu^{(h)} = \nu_1^{(h)} + \dots + \nu^{(h)}_{h-1}$ so that

$$(*) \qquad f_{h\,h+1} = u^{(h)} - \nu^{(h)} \qquad \text{for } h \geqslant 2$$

with

$$(**) \qquad \begin{cases} u^{(h)} \in \Gamma\big(C(\omega_1, \delta_1) \cup \dots \cup C(\omega_{h-1}, \delta_{h-1}) \cup C(\omega_h, T_{h+1}), \mathcal{E}_{S_0}\big) \\ \nu^{(h)} \in \Gamma\big(C(\omega, T_{h+1}), \mathcal{E}_{S_0}\big) \,. \end{cases}$$

If we set $u^{(1)} = f_{12}$ on $C(\omega_1, T_2)$ and $\nu^{(1)} = 0$ then relation $(*)$ and $(**)$ are also valid for $h = 1$.

We set now

$$\begin{aligned} u_1 &= u^{(1)} + u^{(2)} + \dots + u^{(\nu-1)} \\ u_2 &= u^{(2)} + \dots + u^{(\nu-1)} + \nu^{(1)} \\ u_3 &= u^{(3)} + \dots + u^{(\nu-1)} + \nu^{(1)} + \nu^{(2)} \\ &\dots\dots\dots\dots\dots\dots \\ u_{\nu-1} &= u^{(\nu-1)} + \nu^{(1)} + \dots + \nu^{(\nu-2)} \\ u_\nu &= \nu^{(1)} + \dots + \nu^{(\nu-1)} \,. \end{aligned}$$

We have $u_i \in \Gamma(C(\omega_i, T_{i+1}), \mathcal{E}_{S_0})$ for $1 \leqslant i \leqslant \nu$ and from (*) for $1 \leqslant h \leqslant \nu - 1$

$$\begin{aligned} f_{h\,h+1} &= u^{(h)} - v^{(h)} \\ &= u_h - u_{h+1} \, . \end{aligned}$$

This is the condition stated at the end of point (β) that we wanted to prove. Therefore the splitting condition (S) implies that $H^1(\omega, \mathcal{A}_{A_0}) = 0$.

(δ) Conversely let us assume ω non empty open convex such that $H^1(\omega, \mathcal{A}_{A_0}) = 0$. We want to show that ω satisfies the splitting conditions (S).

Let ω_1 be non empty open and convex with $\omega_1 \subset\subset \omega$. Consider an increasing sequence of open convex sets

$$\omega_1 \subset\subset \omega_2 \subset\subset \omega_3 \ldots \subset\subset \omega$$

with $\omega = \bigcup_{j=1}^{\infty} \omega_j$: By T_j with $j = 1, 2, 3, \ldots$ we denote a sequence of real numbers such that

$$T_1 > T_2 > T_3 > \ldots \, , \qquad 0 < T_j < 1/j \, .$$

Let U_j denote any open neighborhood of ω in $\mathbb{R}^N$ such that

$$U_j \supset C(\omega_1, 1/j) \, .$$

If the splitting condition (S) is not valid, for any integer $j \geqslant 1$ we can find T_j such that for any choice of U_j the natural restriction map

$$H^1(C(\omega, T_j) \cup C(\omega_j, 1), \mathcal{E}_{S_0}) \xrightarrow{\alpha} H^1(U_j, \mathcal{E}_{S_0})$$

has a non zero image.

Let

$$\Omega = \bigcup_{j=1}^{\infty} C(\omega_j, T_j) \, .$$

Because ω is convex and by assumption $H^1(\omega, \mathcal{A}_{A_0}) = 0$, by theorem 7 of [4] there exists a neighborhood $\mathcal{B}$ of ω in $\mathbb{R}^N$, contained in Ω such that the natural restriction map

$$H^1(\Omega, \mathcal{E}_{S_0}) \xrightarrow{\beta} H^1(\mathcal{B}, \mathcal{E}_{S_0})$$

has zero image. For any $j \geqslant 1$ we have

$$\Omega \subset C(\omega, T_j) \cup C(\omega_j, 1)$$

and for j large

$$\mathcal{B} \supset C(\omega_1, 1/j)$$

so that $\mathcal{B}$ can be chosen as the open set U_j: But then the map α factors through the map β and we get a contradiction.

SECTION 5

S_0-*Analytic Functionals*

Let $S_0(D)\colon \mathcal{E}^{s_0}(\Omega) \to \mathcal{E}^{s_1}(\Omega)$, for Ω open in $\mathbb{R}^N$ be an elliptic operator on $\mathbb{R}^N$ with constant coefficients. We denote by $\mathcal{E}_{S_0}$ the sheaf of germs of solutions u of the equation $S_0(D)u = 0$.

Let K be a compact set in $\mathbb{R}^N$ and let $\{\Omega_s\}_{s\in\mathbb{N}}$ be a fundamental sequence of open neighborhoods of K. We may assume that $\Omega_s \supset\supset \Omega_{s+1}$ for $s = 1, 2, \ldots$ and that $\Omega_1 \subset\subset \mathbb{R}^N$.

For each s we consider the space $\Gamma(\Omega_s, \mathcal{E}_{S_0})$ endowed with the topology of uniform convergence on compact subsets of Ω_s. Since $S_0(D)$ is an elliptic operator the space $\Gamma(\Omega_s, \mathcal{E}_{S_0})$ with that topology is complete so it is a Fréchet space. Moreover if we embed $\mathbb{R}^N$ into $\mathbb{C}^N$ in the natural way there exists an open neighborhood $\tilde{\Omega}_s$ of Ω_s in $\mathbb{C}^N$ with the following property:

Let $\mathcal{O}_{S_0}$ denote the sheaf of germs of holomorphic functions u in $\mathbb{C}^N$ which satisfy the differential equation $S_0(\partial/\partial z)u = 0$ where $\partial/\partial z = (\partial/\partial z_1, \ldots, \partial/\partial z_N)$ represents the symbol of partial derivative with respect to the holomorphic coordinates in $\mathbb{C}^N$. Consider the Fréchet space $\Gamma(\tilde{\Omega}_s, \mathcal{O}_{S_0})$ the topology being the topology of uniform convergence on compact subsets of $\tilde{\Omega}_s$. Then the natural restriction map $\Gamma(\tilde{\Omega}_s, \mathcal{O}_{S_0}) \to \Gamma(\Omega_s, \mathcal{E}_{S_0})$ is a topological isomorphism.

We may assume that

$$\tilde{\Omega}_s \cap \mathbb{R}^N = \Omega_s\,, \qquad \tilde{\Omega}_s \supset\supset \tilde{\Omega}_{s+1}\,, \qquad \tilde{\Omega}_1 \subset\subset \mathbb{C}^N\,.$$

We have natural restriction maps $\alpha_s\colon \Gamma(\Omega_s, \mathcal{E}_{S_0}) \to \Gamma(\Omega_{s+1}, \mathcal{E}_{S_0})$ and $\beta_s\colon \Gamma(\tilde{\Omega}_s, \mathcal{O}_{S_0}) \to \Gamma(\tilde{\Omega}_{s+1}, \mathcal{O}_{S_0})$ and commutative diagrams of continuous linear maps

$$\begin{array}{ccc}
\Gamma(\Omega_s, \mathcal{E}_{S_0}) & \xrightarrow{\alpha_s} & \Gamma(\Omega_{s+1}, \mathcal{E}_{S_0}) \\
\uparrow \gamma & & \uparrow \gamma \\
\Gamma(\tilde{\Omega}_s, \mathcal{O}_{S_0}) & \xrightarrow{\beta_s} & \Gamma(\tilde{\Omega}_{s+1}, \mathcal{O}_{S_0})\,.
\end{array}$$

Since $\tilde{\Omega}_{s+1} \subset\subset \tilde{\Omega}_s$ the maps β_s, by Vitali theorem, are compact. Therefore also the maps α_s are compact.

We can consider now the space

$$\mathcal{E}_{S_0}(K) = \varinjlim_s \Gamma(\Omega_s, \mathcal{E}_{S_0}) = \varinjlim_s \Gamma(\tilde{\Omega}_s, \mathcal{O}_{S_0})$$

with its topology of inductive limit. This topology is independent from the choice of the sequence $\{\Omega_s\}_{s\in\mathbb{N}}$ and is the topology of a strong dual of a space of Fréchet-Schwartz ([13] p. 135 and 337). Set

$$\phi_{S_0}(K) = \mathcal{E}_{S_0}(K)' = \text{Hom cont.}\,(\mathcal{E}_{S_0}(K), \mathbb{C})$$

With the strong topology this is a space of Fréchet-Schwartz. The topology can be defined as follows:

given $\varepsilon > 0$ let us denote by K_ε the set of points $x \in \mathbb{R}^N$ with

$$\text{dist}\,(x, K) \leqslant \varepsilon;$$

given $\mu \in \phi_{S_0}(K)$ there exists a constant $C = C(\mu, \varepsilon) > 0$ such that

$$|\mu(u)| \leqslant C \sup_{K_\varepsilon} |u|$$

for every $u \in \mathcal{E}_{S_0}(K_\varepsilon)$ (here $|\cdot|$ is any norm on $\mathbb{C}^{s_0}$).

We set

$$\|\mu\|_{K_\varepsilon} = \inf \left\{C > 0 \,\middle|\, |\mu(u)| \leqslant C \sup_{K_\varepsilon} |u| \ \forall u \in \mathcal{E}_{S_0}(K_\varepsilon)\right\}.$$

These are seminorms on $\phi_{S_0}(K)$ and define the topology of $\phi_{S_0}(K)$. Note that if K is a compact and convex set then every $u \in \mathcal{E}_{S_0}(K)$ can be approximated uniformly on K by global functions $v \in \Gamma(\mathbb{R}^N, \mathcal{E}_{S_0})$, note also that for K compact and convex and for any $\varepsilon > 0$ also K_ε is a compact and convex set.

From Hahn-Banach theorem it follows then that the natural restriction map $\Gamma(\mathbb{R}^N, \mathcal{E}_{S_0}) \to \mathcal{E}_{S_0}(K)$ has a dense image so that, for K compact and convex we obtain natural continuous inclusions

$$\lambda_K : \phi_{S_0}(K) \to \mathcal{E}'_{S_0}(\mathbb{R}^N)$$

where, by definition, $\mathcal{E}'_{S_0}(\mathbb{R}^N) = \text{Hom cont}\,(\Gamma(\mathbb{R}^N, \mathcal{E}_{S_0}), \mathbb{C})$.

The elements of $\mathcal{E}'_{S_0}(\mathbb{R}^N)$ could be called *S_0-analytic functionals*. We will say that $\mu \in \mathcal{E}'_{S_0}(\mathbb{R}^N)$ *is carried by the convex compact set* $K \subset \mathbb{R}^N$ if $\mu \in \lambda_K(\phi_{S_0}(K))$ i.e. if for every $\varepsilon > 0$ there exists a constant $c_\varepsilon > 0$ such that

$$|\mu(u)| \leqslant c_\varepsilon \sup_{K_\varepsilon} |u|$$

for every $u \in \Gamma(\mathbb{R}^N, \mathcal{E}_{S_0})$.

As a general notation, given any set $A \subset \mathbb{R}^N$ and u defined in A with values in $\mathbb{C}^{s_0}$, we set

$$\|u\|_A = \sup_A |u|$$

where $|\cdot|$ is any given norm in $\mathbb{C}^{s_0}$.

SECTION 6

The Carrier Condition

Let (1) be a Hilbert complex in $\mathbb{R}^n$ of which we consider an elliptic Cauchy-Kowalewska suspension (2) in $\mathbb{R}^N$.

Let ω be open and convex in $\mathbb{R}^n$. We will say that on ω *the carrier condition is satisfied* with respect to the Cauchy-Kowalewska suspension (2) if the following holds

(C) *For every convex compact $K \subset \omega$ we can find $\delta > 0$ and a convex compact K' with $K \subset K' \subset \omega$ such that given any convex compact set $K'' \subset \omega$, for any S_0-analytic functional $\mu \in \mathcal{E}'_{S_0}(\mathbb{R}^N)$ having the properties*

$$\mu \textit{ has a carrier in } K_\delta$$

$$\mu \textit{ has a carrier in } K''$$

we have also that

$$\mu \textit{ has a carrier in } K' .$$

PROPOSITION 5. *The carrier condition (C) is equivalent to the following condition*

(C_1) *For every convex compact $K \subset \omega$ we can find $\delta > 0$ and a convex compact K' with $K \subset K' \subset \omega$ with the following property: given any convex compact set $K'' \subset \omega$, given T with $0 < T < \delta$ there exist $\varepsilon = \varepsilon(K'', T) > 0$ and $c = c(K'', T) > 0$ such that for any $\mu \in \mathcal{E}'_{S_0}(\mathbb{R}^N)$ satisfying the conditions*

$$|\mu(u)| \leqslant \|u\|_{K_{\delta+\varepsilon}} \qquad \forall u \in \Gamma(\mathbb{R}^N, \mathcal{E}_{S_0})$$

$$|\mu(u)| \leqslant \|u\|_{K''_\varepsilon} \qquad \forall u \in \Gamma(\mathbb{R}^N, \mathcal{E}_{S_0})$$

we also have that

$$|\mu(u)| \leqslant c\|u\|_{K'_T} \qquad \forall u \in \Gamma(\mathbb{R}^N, \mathcal{E}_{S_0}) .$$

PROOF. $(C_1) \Rightarrow (C)$. Given K we determine $\delta > 0$ and K' according to condition (C_1). Let K'' convex compact of ω be given and let $\mu \in \mathcal{E}'_{S_0}(\mathbb{R}^N)$ be such that

$$(*) \qquad \begin{aligned} &\mu \text{ has a carrier in } K_\delta \\ &\mu \text{ has a carrier in } K'' . \end{aligned}$$

Given $T > 0$, with $0 < T < \delta$, we determine $\varepsilon = \varepsilon(K'', T) > 0$ and $c = c(K'', T) > 0$ by condition (C_1). Because of the assumption $(*)$ there exist constants $c_{1\varepsilon} > 0$ $c_{2\varepsilon} > 0$ such that

$$\begin{aligned} |\mu(u)| &\leqslant c_{1\varepsilon}\|u\|_{K_{\delta+\varepsilon}} \\ |\mu(u)| &\leqslant c_{2\varepsilon}\|u\|_{K''_\varepsilon} \end{aligned} \qquad \text{for every } u \in \Gamma(\mathbb{R}^N, \mathcal{E}_{S_0}) .$$

Then by condition (C_1) we also have

$$|\mu(u)| \leqslant \sup(c_{1\varepsilon}, c_{2\varepsilon}) c \|u\|_{K'_T} .$$

This shows that K' is also a carrier of μ. Thus condition (C) holds.

$(C) \Rightarrow (C_1)$. Given K we determine $\delta > 0$ and K' according to condition (C).

Let K'' be a given compact convex subset of ω. Consider the spaces:

$$G = \{(\mu_1, \mu_2, \mu_3) \in \phi_{S_0}(K_\delta) \times \phi_{S_0}(K'') \times \phi_{S_0}(K') | \lambda_{K_\delta}(\mu_1) = \lambda_{K''}(\mu_2) = \lambda_{K'}(\mu_3)\},$$

$$F = \{(\mu_1, \mu_2) \in \phi_{S_0}(K_\delta) \times \phi_{S_0}(K'') | \lambda_{K_\delta}(\mu_1) = \lambda_{K''}(\mu_2)\} .$$

These are spaces of Fréchet-Schwartz because the spaces $\phi_{S_0}(K_\delta)$, $\phi_{S_0}(K'')$ and $\phi_{S_0}(K')$ are such and because the maps $\lambda_{K_\delta}, \lambda_{K''}, \lambda_{K'}$ are continuous.

Moreover the natural projection

$$\pi\colon G \to F$$

is continuous.

If condition (C) holds, then π is surjective. Therefore π is open by Banach's theorem.

It follows that for every $T > 0$ (with $0 < T < \delta$) we can find $\varepsilon = \varepsilon(T, K'') > 0$ and $\varrho = \varrho(T, K'') > 0$ such that

$$\{(\mu_1, \mu_2) \in F | \|\mu_1\|_{K_{\delta+\varepsilon}} < \varrho, \|\mu_2\|_{K''_\varepsilon} < \varrho\} \subset \pi\{(\mu_1, \mu_2, \mu_3) \in G | \|\mu_3\|_{K'_T} < 1\} .$$

Therefore, for every $\mu \in \mathcal{E}'_{S_0}(\mathbb{R}^n)$ with

$$|\mu(u)| \leqslant \|u\|_{K_{\delta+\varepsilon}} \qquad \forall u \in \Gamma(\mathbb{R}^N, \mathcal{E}_{S_0})$$

$$|\mu(u)| \leqslant \|u\|_{K''_\varepsilon} \qquad \forall u \in \Gamma(\mathbb{R}^N, \mathcal{E}_{S_0})$$

we also have

$$|\mu(u)| \leqslant \frac{1}{\varrho} \|u\|_{K'_T} \qquad \forall u \in \Gamma(\mathbb{R}^N, \mathcal{E}_{S_0}) .$$

This shows that condition (C_1) holds with $\varepsilon = \varepsilon(T, K'')$ and $c = 1/\varrho(T, K'')$.

PROPOSITION 6. *Condition* (C_1) *can be reformulated in the following equivalent form*

(C_2) *For every compact convex set* $L \subset \omega$ *we can find* $\sigma > 0$ *and a convex compact set* L' *with* $L \subset L' \subset \omega$ *with the following property*

given any compact convex set $L'' \subset \omega$, *given any* S *with* $0 < S < \sigma$ *there exist* $\eta = \eta(L'', S) > 0$ *and* $k = k(L'', S) > 0$ *such that*

for any $\mu \in \mathcal{E}'_{S_0}(\mathbb{R}^N)$ *satisfying the conditions*

$$|\mu(u)| \leqslant \|u\|_{C(L,\sigma)} \qquad \forall u \in \Gamma(\mathbb{R}^N, \mathcal{E}_{S_0})$$

$$|\mu(u)| \leqslant \|u\|_{C(L'',\eta)} \qquad \forall u \in \Gamma(\mathbb{R}^N, \mathcal{E}_{S_0})$$

we also have that

$$|\mu(u)| \leqslant k \|u\|_{C(L',S)} \qquad \forall u \in \Gamma(\mathbb{R}^N, \mathcal{E}_{S_0}) .$$

PROOF. $(C_2) \Rightarrow (C_1)$. Let the convex compact set $K \subset \omega$ be given and let $0 < \lambda < \operatorname{dist}(K, \partial\omega)$. Let $L = K_\lambda \cap \mathbb{R}^n$; this is a convex compact subset of ω. We can then by condition (C_2) determine $\sigma > 0$ and $L' \subset \omega$ convex and compact with the properties specified by (C_2). We set $K' = L'$ and choose $\delta = \inf(\sigma/2, \lambda/2)$. Let now the compact convex set $K'' \subset \omega$ be given and let $0 < \lambda' < \operatorname{dist}(K'', \partial\omega)$. We choose $L'' = K''_{\lambda'} \cap \mathbb{R}^n$ as a convex compact subset of ω. We fix T with $0 < T < \delta$ and set $\varepsilon = \inf(\sigma/2, \eta(L'', T), \lambda', \lambda/2)$ $c = k(L'', T)$.

We then have

$$K_{\delta+\varepsilon} \subset C(K_\lambda \cap \mathbb{R}^n, \sigma) \subset C(L, \sigma)$$

$$K''_\varepsilon \subset C(K''_{\lambda'} \cap \mathbb{R}^n, \eta) \subset C(L'', \eta)$$

and

$$K'_T \supset C(L', T)$$

as $K' = L'$. From these inclusions the implication $(C_2) \Rightarrow (C_1)$ is straightforward.

$(C_1) \Rightarrow (C_2)$. Let the convex compact set $L \subset \omega$ be given.

Put $K = L$. We determine $\delta > 0$ and K' convex compact in ω by (C_1). Let $0 < \tau < \operatorname{dist}(K', \partial\omega)$ and set $\sigma = \inf(\delta, \tau)$. We then choose $L' = K'_\tau \cap \mathbb{R}^n$. Let L'' be a given convex compact subset of ω and set $K'' = L''$. We choose S with $0 < S < \sigma$ and set $\eta = \varepsilon(L'', S)$, $k = c(L'', S)$.

We have the inclusions

$$C(L, \sigma) \subset K_{\delta+\varepsilon}$$

$$C(L'', \eta) \subset K''_\varepsilon$$

$$C(L', S) \supset K'_S .$$

From these the implication $(C_1) \Rightarrow (C_2)$ follows.

SECTION 7

Characterization of the Carrier Condition

a) Let $S_0(D)\colon \mathcal{E}^{s_0}(\Omega) \to \mathcal{E}^{s_1}(\Omega)$ for Ω open in $\mathbb{R}^N$ be an elliptic operator with constant coefficients defined on $\mathbb{R}^N$. Let $\mathcal{E}_{S_0}$ denote the sheaf of germs u of solutions of the equation $S_0(D)u = 0$. Given Ω open in $\mathbb{R}^N$ the space $\Gamma(\Omega, \mathcal{E}_{S_0})$ with the topology of uniform convergence on compact sets K of Ω is a complete and thus a Fréchet space, the topology being defined by the seminorms $\|u\|_K = \sup_K |u|$ where $|\cdot|$ is any given norm on $\mathbb{C}^{s_0}$.

The space $\mathcal{E}^{s_0}(\Omega)$ being topologized with the Schwartz topology (uniform convergence on compact subsets of Ω of the functions $u \in \mathcal{E}^{s_0}(\Omega)$ and all their partial derivatives), we denote by $\mathcal{E}'^{s_0}(\Omega)$ the space Hom cont. $(\mathcal{E}^{s_0}(\Omega), \mathbb{C})$ of distributions with compact support in Ω.

LEMMA 3. *Let $\Omega, \Omega_1, \Omega_2$ be open sets in $\mathbb{R}^N$ with Ω connected. Assume that for $u \in \Gamma(\Omega, \mathcal{E}_{S_0})$ we can find $u_1 \in \Gamma(\Omega_1, \mathcal{E}_{S_0})$ and $u_2 \in \Gamma(\Omega_2, \mathcal{E}_{S_0})$ such that*

$$u = u_1 - u_2 \quad \text{on } \Omega \cap \Omega_1 \cap \Omega_2 .$$

Then given compact sets $K_1 \subset \Omega_1$ and $K_2 \subset \Omega_2$ we can find a compact set $K \subset \Omega$ and a constant $c > 0$ such that for every $u \in \Gamma(\Omega, \mathcal{E}_{S_0})$ we can choose $u_1 \in \Gamma(\Omega_1, \mathcal{E}_{S_0})$ and $u_2 \in \Gamma(\Omega_2, \mathcal{E}_{S_0})$ to have

$$u = u_1 - u_2 \quad \text{on } \Omega \cap \Omega_1 \cap \Omega_2$$

$$\|u_1\|_{K_1} + \|u_2\|_{K_2} \leqslant c\|u\|_K .$$

Moreover any distribution $\mu \in \mathcal{E}'^{s_0}(\Omega \cap \Omega_1 \cap \Omega_2)$ which satisfies the conditions

$$|\mu(u)| \leqslant \|u\|_{K_1} \qquad \forall u \in \Gamma(\Omega_1, \mathcal{E}_{S_0})$$

$$|\mu(u)| \leqslant \|u\|_{K_2} \qquad \forall u \in \Gamma(\Omega_2, \mathcal{E}_{S_0})$$

also satisfies the condition

$$|\mu(u)| \leqslant c\|u\|_K \qquad \forall u \in \Gamma(\Omega, \mathcal{E}_{S_0}) .$$

PROOF. Let

$$F = \{(u, u_1, u_2) \in$$

$$\in \Gamma(\Omega, \mathcal{E}_{S_0}) \times \Gamma(\Omega_1, \mathcal{E}_{S_0}) \times \Gamma(\Omega_2, \mathcal{E}_{S_0}) | u = u_1 - u_2 \text{ on } \Omega \cap \Omega_1 \cap \Omega_2\} .$$

This is a closed subspace of the Fréchet space $\Gamma(\Omega, \mathcal{E}_{S_0}) \times \Gamma(\Omega_1, \mathcal{E}_{S_0}) \times \Gamma(\Omega_2, \mathcal{E}_{S_0})$ and thus it is a Fréchet space.

Let

$$\pi: F \to \Gamma(\Omega, \mathcal{E}_{S_0})$$

be the continuous linear map induced on F by the projection of $\Gamma(\Omega, \mathcal{E}_{S_0}) \times \times \Gamma(\Omega_1, \mathcal{E}_{S_0}) \times \Gamma(\Omega_2, \mathcal{E}_{S_0})$ on its first factor.

By the assumption this map is surjective and therefore open by Banach theorem. Then there exists a compact set $K \subset \Omega$ and an $\varepsilon > 0$ such that

$$\pi\{(u, u_1, u_2) \in F | \|u_1\|_{K_1} + \|u_2\|_{K_2} < 1\} \subset \{u \in \Gamma(\Omega, \mathcal{E}_{S_0}) | \|u\|_K < \varepsilon\} .$$

It is not restrictive, enlarging K, to assume that $\mathring{K} \neq \emptyset$.

This shows that every $u \in \Gamma(\Omega, \mathcal{E}_{S_0})$ can be written as

$$u = u_1 - u_2 \quad \text{with } u_1 \in \Gamma(\Omega_1, \mathcal{E}_{S_0}) , \quad u_2 \in \Gamma(\Omega_2, \mathcal{E}_{S_0})$$

such that

$$\|u_1\|_{K_2} + \|u_2\|_{K_1} \leqslant \frac{2}{\varepsilon} \|u\|_K .$$

Indeed if $u \neq 0$ then $\|u\|_K \neq 0$ as u is analytic, $\mathring{K} \neq \emptyset$ and Ω is connected. Thus $(\varepsilon/2)(u/\|u\|_K) \in \{u \in \Gamma(\Omega, \mathcal{E}_{S_0}) | \|u\|_K < \varepsilon\}$. This completes the proof of the first statement of the lemma taking $c = 2/\varepsilon$.

Let now $\mu \in \mathcal{E}'^{s_0}(\Omega \cap \Omega_1 \cap \Omega_2)$ with the properties mentioned in the lemma; then for $u \in \Gamma(\Omega, \mathcal{E}_{S_0})$ we have

$$\begin{aligned} |\mu(u)| &= |\mu(u_1 - u_2)| \leqslant |\mu(u_1)| + |\mu(u_2)| \\ &\leqslant \|u_1\|_{K_1} + \|u_2\|_{K_2} \\ &\leqslant c\|u\|_K \end{aligned}$$

if u_1 and u_2 are chosen as before.

We define for any open set $\Omega \subset \mathbb{R}^N$

$$B(\Omega, \mathcal{E}_{S_0}) = \{u \in \Gamma(\Omega, \mathcal{E}_{S_0}) | \|u\|_\Omega < \infty\} .$$

LEMMA 5. *Let Ω, Ω_1, Ω_2 be open sets in* $\mathbb{R}^N$ *with* $\Omega \cap \Omega_1 \cap \Omega_2$ *non empty and connected. Let* U *be a non empty open subset of* $\Omega \cap \Omega_1 \cap \Omega_2$.

We assume that there exists a constant $c > 0$ *with the following property:*

every $\mu \in \mathcal{E}'^{s_0}(U)$ *which satisfies the conditions*

$$|\mu(u)| \leqslant \|u\|_{\Omega_1} \qquad \forall u \in B(\Omega_1, \mathcal{E}_{S_0})$$

$$|\mu(u)| \leqslant \|u\|_{\Omega_2} \qquad \forall u \in B(\Omega_2, \mathcal{E}_{S_0})$$

also satisfies the condition

$$|\mu(u)| \leqslant c\|u\|_{\Omega} \qquad \forall u \in B(\Omega, \mathcal{E}_{S_0})\,.$$

Then for every $u \in B(\Omega, \mathcal{E}_{S_0})$ *we can find* $u_1 \in B(\Omega_1, \mathcal{E}_{S_0})$ *and* $u_2 \in B(\Omega_2, \mathcal{E}_{S_0})$ *such that*

$$u = u_1 - u_2 \qquad \text{on } \Omega \cap \Omega_1 \cap \Omega_2\,,$$

$$\|u_1\|_{\Omega_1} + \|u_2\|_{\Omega_2} \leqslant c\|u\|_{\Omega}\,.$$

PROOF. Set

$$A = \{u \in \mathcal{E}^{s_0}(U) | u = u_1 - u_2 \text{ on } U,$$
$$\text{with } u_1 \in B(\Omega_1, \mathcal{E}_{S_0}),\, u_2 \in B(\Omega_2, \mathcal{E}_{S_0}),\, \|u_1\|_{\Omega_1} + \|u_2\|_{\Omega_2} \leqslant 1\}\,.$$

The set A is convex and disked. Moreover A is compact.

Indeed let $u_n = u_{1n} - u_{2n}$ be a subsequence of A. Because $\|u_{1n}\|_{\Omega_1} + \|u_{2n}\|_{\Omega_2} \leqslant 1$, and as $S_0(D)$ is elliptic, we can find subsequences $\{u_{1n_h}\}$ and $\{u_{2n_h}\}$ which converge uniformly on any compact subset of Ω_1 and Ω_2 respectively. This by virtue of Vitali's theorem. Let $u_1 = \lim_h u_{1n_h}$, $u_2 = \lim_h u_{2n_h}$. For any choice of compact sets $K_1 \subset \Omega_1$ and $K_2 \subset \Omega_2$ we have $\|u_1\|_{K_1} + \|u_2\|_{K_2} \leqslant 1$. Therefore also $\|u_1\|_{\Omega_1} + \|u_2\|_{\Omega_2} \leqslant 1$. Set on U

$$u = \lim (u_{1n_h} - u_{2n_h}) = \lim u_{n_h}\,.$$

On every compact subset $K \subset U$ u_{n_h} converges uniformly to u with all partial derivatives. Therefore u_{n_h} converges to u in the topology of $\mathcal{E}^{s_0}(U)$. Hence A is compact.

Set

$$A^0 = \{\mu \in \mathcal{E}'^{s_0}(U) | |\mu(u)| \leqslant 1 \;\; \forall u \in A\}\,.$$

We have

$$A = A^{00} = \{u \in \mathcal{E}^{s_0}(U) \mid |\mu(u)| \leqslant 1 \ \ \forall \mu \in A^0\} .$$

Let now $\mu \in A^0$. For every $u \in B(\Omega_i, \mathcal{E}_{S_0})$, $u \neq 0$, we have that $u/\|u\|_{\Omega_i} \in A$, $i = 1, 2$. Therefore we have

$$|\mu(u)| \leqslant \|u\|_{\Omega_1} \qquad \forall u \in B(\Omega_1, \mathcal{E}_{S_0})$$
$$|\mu(u)| \leqslant \|u\|_{\Omega_2} \qquad \forall u \in B(\Omega_2, \mathcal{E}_{S_0}) .$$

By the assumption then every $\mu \in A^0$ satisfies also the condition

$$|\mu(u)| \leqslant c\|u\|_{\Omega} \qquad \forall u \in B(\Omega, \mathcal{E}_{S_0}) .$$

But for $u \neq 0$ in $B(\Omega, \mathcal{E}_{S_0})$ this means that $u/c\|u\|_{\Omega} \in A^{00} = A$ i.e.

(*) $\qquad u = u_1 - u_2 \quad$ on U, $\qquad$ for every $u \in B(\Omega, \mathcal{E}_{S_0})$

with

$$\|u_1\|_{\Omega_1} + \|u_2\|_{\Omega_2} \leqslant c\|u\|_{\Omega}$$

and $u_1 \in B(\Omega_1, \mathcal{E}_{S_0})$, $u_2 \in B(\Omega_2, \mathcal{E}_{S_0})$. As $\Omega \cap \Omega_1 \cap \Omega_2$ is connected and $u - u_1 + u_2$ is analytic there and zero on $U \neq \emptyset$ we conclude that (*) holds in the whole of $\Omega \cap \Omega_1 \cap \Omega_2$.

THEOREM 2. *Let ω be an open convex set in $\mathbb{R}^n$.*

The set ω satisfies the carrier condition (C) with respect to the suspension (2) of the complex (1) if and only if ω is analytically convex with respect to the complex (1) i,e. if and only if

$$H^1(\omega, \mathcal{A}_{A_0}) = 0 .$$

PROOF. By virtue of theorem 1 it is enough to show that the carrier condition (C) is equivalent to the splitting condition (S). We will take the carrier condition under the form (C_2) of the proposition 6.

$(S) \Rightarrow (C_2)$. Let $K \subset \omega$ be a compact and convex set. We choose ω_1 open and convex with $K \subset \omega_1 \subset\subset \omega$. By condition (S) we can find $\delta_1 > 0$ and an open convex set ω_2 with $\omega_1 \subset \omega_2 \subset\subset \omega$ such that for any T with $0 < T < \delta_1$ there exists an open neighborhood $\Omega(T) \subset C(\omega, T)$ of ω in $\mathbb{R}^N$ with the property:

For any $u \in \Gamma(C(\omega_2, T), \mathcal{E}_{S_0})$ we can find

$$u_1 \in \Gamma(C(\omega_1, \delta_1), \mathcal{E}_{S_0}) \quad \text{and} \quad u_2 \in \Gamma(\Omega(T), \mathcal{E}_{S_0})$$

so that

$$u = u_1 - u_2 \quad \text{on } C(\omega_1, \delta_1) \cap \Omega(T).$$

We choose K' compact convex with $\omega_2 \subset K' \subset \omega$.

We now apply the corollary to proposition 0 with $\Omega_1 = C(\omega_1, \delta_1)$ and $\Omega_2 = \Omega(T)$. We determine, according to that corollary, the open sets $\Omega_1' = C(\omega_1, \delta_1)'$ and $\Omega_2' = \Omega(T)'$ with $\Omega_1' \subset \Omega_1$, $\Omega_2' \subset \Omega_2$, so that any $f \in \Gamma(C(\omega_1, \delta_1) \cup \Omega(T), \mathcal{E}_{S_0})$ can be approximated uniformly on compact subsets of $C(\omega_1, \delta_1)' \cup \Omega(T)'$ by global functions $g \in \Gamma(\mathbb{R}^N, \mathcal{E}_{S_0})$.

We select $\delta > 0$ so that

$$C(K, \delta) \subset C(\omega_1, \delta_1)'$$

and we choose T with $0 < T < \delta$.

Given a convex compact set K'' we determine $\varepsilon > 0$ so that

$$C(K'', \varepsilon) \subset \Omega(T)' .$$

By lemma 4 given the compact sets $C(K, \delta) \subset C(\omega_1, \delta_1)$ and $C(K'', \varepsilon) \subset \Omega(T)$ there exists a compact set $L \subset C(\omega_2, T)$ and a positive constant $c > 0$ so that for any $u \in \Gamma(C(\omega_2, T), \mathcal{E}_{S_0})$, $u_1 \in \Gamma(C(\omega_1, \delta_1), \mathcal{E}_{S_0})$ and $u_2 \in \Gamma(\Omega(T), \mathcal{E}_{S_0})$ can be so chosen that $u = u_1 - u_2$ on $C(\omega_1, \delta_1) \cap \Omega(T)$ and so that

$$(*) \qquad \|u_1\|_{C(K,\delta)} + \|u_2\|_{C(K'',\varepsilon)} \leqslant c\|u\|_L .$$

Let now $\mu \in \mathcal{E}_{S_0}'(\mathbb{R}^N)$ and assume that

$$|\mu(u)| \leqslant \|u\|_{C(K,\delta)}$$

$$|\mu(u)| \leqslant \|u\|_{C(K'',\varepsilon)}$$

for any $u \in \Gamma(\mathbb{R}^N, \mathcal{E}_{S_0})$. We want to show that we have also

$$|\mu(u)| \leqslant (c+2)\|u\|_{C(K',T)} .$$

Let $u \in \Gamma(\mathbb{R}^N, \mathcal{E}_{S_0})$ be given with $u \neq 0$. We determine u_1 and u_2 as above with $u = u_1 - u_2$ on $C(\omega_1, \delta_1) \cap \Omega(T)$ and with the estimate $(*)$.

Now $u_1 = u - u_2$ can be defined by the right hand side on $\Omega(T)$ so that u_1 is defined on $C(\omega_1, \delta_1) \cup \Omega(T)$.

By the above remark we can find $g \in \Gamma(\mathbb{R}^N, \mathcal{E}_{S_0})$ so that

$$(**) \qquad \|u_1 - g\|_{C(K,\delta) \cup C(K'',\varepsilon)} \leqslant \|u\|_{C(\omega_2,T)} .$$

We can then write

$$u = g + (u_1 - g - u_2) = g - h$$

and $h = g - u \in \Gamma(\mathbb{R}^N, \mathcal{E}_{S_0})$. We have

$$\begin{aligned}
|\mu(u)| &= |\mu(g) - \mu(h)| \\
&\leqslant |\mu(g)| + |\mu(h)| \\
&\leqslant \|g\|_{C(K,\delta)} + \|h\|_{C(K'',\varepsilon)} \qquad \text{(by the assumptions)} \\
&\leqslant \|g - u_1\|_{C(K,\delta)} + \|u_1\|_{C(K,\delta)} \\
&\quad + \|u_1 - g\|_{C(K'',\varepsilon)} + \|u_2\|_{C(K'',\varepsilon)} \\
&\leqslant (c+2)\|u\|_{C(\omega_2,T)} \qquad \text{(because of } (*) \text{ and } (**)) \\
&\leqslant (c+2)\|u\|_{C(K',T)} \qquad (\text{as } \omega_2 \subset K') .
\end{aligned}$$

This proves our contention and therefore condition (C_2) holds with $L = K$, $L' = K'$, $\sigma = \delta$, $S = T$, $L'' = K''$ and $\eta = \varepsilon$.

$(C_2) \Rightarrow (S)$. Let $\omega_1 \subset\subset \omega$ be a given non empty open convex subset of ω. We choose K compact and convex with $\omega_1 \subset K \subset \omega$. By condition (C_2) we determine then a compact convex set $K' \subset \omega$ and $\delta > 0$ with $K \subset K'$.

We choose ω_2 open convex and ω'' open convex such that

$$K' \subset \omega_2 \subset\subset \omega'' \subset\subset \omega .$$

We finally choose K'' compact convex with $\omega'' \subset K'' \subset \omega$.

By the validity of condition (C_2) for any T with $0 < T < \delta$ we can determine $\varepsilon = \varepsilon(K'', T/2) > 0$ and $c = c(K'', T/2) > 0$ such that:

any S_0-analytic functional $\mu \in \mathcal{E}'_{S_0}(\mathbb{R}^N)$ which verifies the conditions

$$|\mu(u)| \leqslant \|u\|_{C(K,\delta)} \qquad \forall u \in \Gamma(\mathbb{R}^N, \mathcal{E}_{S_0})$$

$$|\mu(u)| \leqslant \|u\|_{C(K'',\varepsilon)} \qquad \forall u \in \Gamma(\mathbb{R}^N, \mathcal{E}_{S_0})$$

also satisfies the condition

$$|\mu(u)| \leqslant c\|u\|_{C(K',T/2)} \qquad \forall u \in \Gamma(\mathbb{R}^N, \mathcal{E}_{S_0}) .$$

We claim then that the natural restriction map

$$r\colon H^1\big(C(\omega_2, \delta) \cup C(\omega'', T), \mathcal{E}_{S_0}\big) \to H^1\big(C(\omega_1, \delta) \cup C(\omega'', \varepsilon), \mathcal{E}_{S_0}\big)$$

has zero image.

Choose α open convex with $K' \subset \alpha \subset\subset \omega_2$, and let U be open non empty with $U \subset C(\omega_1, \delta) \cap C(\omega'', \varepsilon) \cap C(\alpha, T/3)$. Let $\mu \in \mathcal{E}'^{s_0}(U)$ and assume that

$$|\mu(u)| \leqslant \|u\|_{C(\omega_1,\delta)} \qquad \forall u \in B\big(C(\omega_1, \delta), \mathcal{E}_{S_0}\big)$$
$$|\mu(u)| \leqslant \|u\|_{C(\omega'',\varepsilon)} \qquad \forall u \in B\big(C(\omega'', \varepsilon), \mathcal{E}_{S_0}\big)\,.$$

Then $|\mu(u)| \leqslant \|u\|_{C(K,\delta)}$ and $|\mu(u)| \leqslant \|u\|_{C(K'',\varepsilon)}$ for every $u \in \Gamma(\mathbb{R}^N, \mathcal{E}_{S_0})$.

Therefore $|\mu(u)| \leqslant C\|u\|_{C(K',T/2)}$ for every $u \in \Gamma(\mathbb{R}^N, \mathcal{E}_{S_0})$.

Let now $u \in B\big(C(\alpha, \frac{3}{4}T), \mathcal{E}_{S_0}\big)$. We can find a sequence $\{u_\nu\}_{\nu\in\mathbb{N}}$ in $\Gamma(\mathbb{R}^N, \mathcal{E}_{S_0})$ such that $\|u - u_\nu\|_{C(K',T/2)} \to 0$ for $\nu \to \infty$.

We have for $\nu \to \infty$ $\mu(u_\nu) \to \mu(u)$ as $\mu \in \mathcal{E}'^{s_0}(U)$ and $U \subset C(\omega_1, T/3) \subset C(K', T/2)$. Also $|\mu(u_\nu)| \leqslant C\|u_\nu\|_{C(K',T/2)}$. Therefore passing to the limit for $\nu \to \infty$ we get

$$|\mu(u)| \leqslant c\|u\|_{C(\alpha,\frac{3}{4}T)} \qquad \forall u \in B\big(C(\alpha, \tfrac{3}{4}T), \mathcal{E}_{S_0}\big)\,.$$

By lemma 5 any $u \in B\big(C(\alpha, \frac{3}{4}T), \mathcal{E}_{S_0}\big)$ can be written as $u = u_1 - u_2$ with $u_1 \in B\big(C(\omega_1, \delta), \mathcal{E}_{S_0}\big)$ and $u_2 \in B\big(C(\omega'', \varepsilon), \mathcal{E}_{S_0}\big)$ on

$$C(\omega_1, \delta) \cap C(\omega'', \varepsilon) \cap C(\alpha, \tfrac{3}{4}T)\,.$$

Let now $u \in \Gamma\big(C(\omega_2, T), \mathcal{E}_{S_0}\big)$ be given. Then $u|C(\alpha, \frac{3}{4}T)$ defines an element $u \in B\big(C(\alpha, \frac{3}{4}T), \mathcal{E}_{S_0}\big)$. By the conclusion, above stated, we can find

$$u_1 \in \Gamma\big(C(\omega_1, \delta), \mathcal{E}_{S_0}\big) \quad \text{and } u_2 \in \Gamma\big(C(\omega'', \varepsilon), \mathcal{E}_{S_0}\big)$$

such that

$$(*) \qquad u = u_1 - u_2$$

on $C(\omega_1, \delta) \cap C(\omega'', \varepsilon) \cap C(\alpha, \frac{3}{4}T)$. But u, u_1, u_2 are analytic. Therefore the condition $(*)$ must hold on $C(\omega_1, \delta) \cap C(\omega'', \varepsilon) \cap C(\omega_2, T) = C(\omega_1, T) \cap C(\omega'', \varepsilon)$.

This proves that the map r has zero image as we claimed.

Consider now an elliptic and Cauchy-Kowalewska suspension (3) of the

Hilbert complex (1) to $\mathbb{R}^{n+1}$ ($H = n + 1$). From the remark following proposition 2 we deduce the following statement:

for any choice of non empty convex sets ω_1' *and* ω_2' *with*

$$\omega_1' \subset\subset \omega_1 \quad \textit{and} \quad \omega_2 \subset\subset \omega_2' \subset\subset \omega$$

we can find $\sigma > 0$ *and for any* T' *with* $0 < T' < \sigma$ *an open neighborhood* $\mathfrak{U}(T') \subset C(\omega'', T')$ *of* ω'' *in* $\mathbb{R}^{n+1}$ *such that the natural restriction map*

$$H^1(C_H(\omega'', T') \cup C_H(\omega_2', \sigma), \mathcal{E}_{L_0}) \to H^1(\mathfrak{U}(T') \cup C_H(\omega_1', \sigma), \mathcal{E}_{L_0})$$

has zero image.

But then from the remark following the proof of proposition 3 we deduce that also the natural restriction map

$$H^1(C_H(\omega'', T') \cup C_H(\omega_2', \sigma), \mathcal{E}_{L_0}) \to H^1(C_H(\omega'', T') \cup C_H(\omega_1', \sigma), \mathcal{E}_{L_0})$$

has zero image.

We now choose $\omega'' = \omega_\nu''$ $\nu = 1, 2, \dots$ as an increasing sequence of open convex sets ω_ν'' which tend to ω:

$$\omega_1'' \subset \omega_2'' \subset \omega_3'' \dots \subset \omega_\nu'' \subset\subset \omega\,, \qquad \bigcup_{\nu=1}^{\infty} \omega_\nu'' = \omega\,.$$

By the previous remark it follows that for every ν the natural restriction map

$$H^1(C_H(\omega, T') \cup C_H(\omega_2', \sigma), \mathcal{E}_{L_0}) \to H^1(C_H(\omega_\nu'', T') \cup C_H(\omega_1', \sigma), \mathcal{E}_{L_0})$$

has zero image

Also for every c the restriction map

$$\Gamma(\mathbb{R}^{n+1}, \mathcal{E}_{L_0}) \to \Gamma(C_H(\omega_\nu'', T') \cup C_H(\omega', \sigma), \mathcal{E}_{L_0})$$

has a dense image by theorem 5 of [4]. Therefore we can apply proposition 14 of [4] and conclude that the natural restriction map

$$H^1(C_H(\omega, T') \cup C_H(\omega_2', \sigma), \mathcal{E}_{L_0}) \to H^1(C_H(\omega, T') \cup C_H(\omega_1', \sigma), \mathcal{E}_{L_0})$$

has zero image.

But ω_1' can be any open non empty convex set $\omega_1' \subset\subset \omega$ (as we can choose ω_1 after the choice of ω_1' with $\omega_1' \subset\subset \omega_1 \subset\subset \omega$). It follows then that the splitting condition (S') holds for the suspension (3) of (1).

Then, by proposition 2, for any elliptic and Cauchy-Kowalewska suspension (2) of the complex (1) the splitting condition (S) also holds.

SECTION 8

The $\bar{\partial}$-Suspension. Some Algebraic Preliminaries

a) We consider in $\mathbb{R}^n$ the Hilbert complex

$$(1)\qquad \mathcal{E}^{p_0}(\omega) \xrightarrow{A_0(D)} \mathcal{E}^{p_1}(\omega) \xrightarrow{A_1(D)} \mathcal{E}^{p_2}(\omega) \to \dots$$

where $D = D_x = (\partial/\partial x_1, \dots, \partial/\partial x_n)$, $x_1, \dots, x_n$ being cartesian coordinates in $\mathbb{R}^n$ and ω denotes an open set in $\mathbb{R}^n$.

The complex (1) is obtained from a Hilbert resolution of a finitely generated module M over the ring $\mathfrak{P}_n = \mathbb{C}[\xi_1, \dots, \xi_n]$ of polynomials in n variables:

$$(1^*)\qquad 0 \leftarrow M \leftarrow \mathfrak{P}_n^{p_0} \xleftarrow{{}^tA_0(\xi)} \mathfrak{P}_n^{p_1} \xleftarrow{{}^tA_1(\xi)} \mathfrak{P}_n^{p_2} \leftarrow \dots .$$

We consider the space $\mathbb{R}^n$ where ξ are coordinates as imbedded in $\mathbb{R}^{2n} = \mathbb{C}^n$ where $\xi_1 + i\eta_1, \dots, \xi_n + i\eta_n$ are complex coordinates, and similarly the space $\mathbb{R}^n$ where x are coordinates will be considered imbedded in the space $\mathbb{C}^n$ where $z_1 = x_1 + iy_1, \dots, z_n = x_n + iy_n$ are complex coordinates.

We consider from now on only the $\bar{\partial}$-suspension to $\mathbb{C}^n$ of the Hilbert complex (1). This is an elliptic Cauchy-Kowalewska suspension in $\mathbb{C}^n$ of the complex (1) which begins with the operator (cf. [4] n. 4 α) β))

$$S_0(D) = \bar{\partial} \oplus A_0(D_x)$$

$$\mathcal{E}^{p_0}(\Omega) \begin{array}{l} \xrightarrow{\ \bar{\partial}\ } (\mathcal{E}^{01}(\Omega))^{p_0} \\ \qquad\quad \oplus \dots \\ \xrightarrow{A_0(D_x)} \mathcal{E}^{p_1}(\Omega) \end{array}$$

where $\mathcal{E}^{01}(\Omega)$ denotes C^∞ forms of type $(0, 1)$ on the open set $\Omega \subset \mathbb{C}^n$.

The sheaf $\mathcal{E}_{S_0}$ is therefore isomorphic with the sheaf $\mathcal{O}_{A_0}$ of germs of holomorphic functions u with values in $\mathbb{C}^{p_0}$ satisfying the equation

$$A_0\left(\frac{\partial}{\partial z}\right)u = 0 .$$

This $\bar{\partial}$-suspension is the complex that corresponds to the tensor product (over $\mathcal{T}_n$) of the complex (1*) and the complex

$$(2^*) \qquad 0 \leftarrow N \leftarrow \mathcal{T}_{2n}^{(n)} \xleftarrow{\wedge\alpha} \mathcal{T}_{2n}^{(n-1)} \xleftarrow{\wedge\alpha} \ldots \leftarrow \mathcal{T}_{2n}^{(1)} \xleftarrow{\wedge\alpha} \mathcal{T}_{2n}^{(0)} \leftarrow 0$$

where

$$\alpha = (\xi_1 + i\eta_1)\,dt_1 + \ldots. + (\xi_n + i\eta_n)\,dt_n$$

and where $\mathcal{T}_{2n}^{(i)}$ denotes the space of exterior differential forms of degree i in $dt_1, \ldots, dt_n$ with coefficients in $\mathcal{T}_{2n} = \mathbb{C}[\xi_1, \ldots, \xi_n, \eta_1, \ldots, \eta_n]$.

The complex (2*) is the Hilbert resolution that corresponds to the Dolbeault complex in $\mathbb{C}^n$.

The module N is isomorphic with

$$N \simeq \mathbb{C}[\xi_1, \ldots, \xi_n] = \mathcal{T}_n = \frac{\mathbb{C}[\xi_1, \ldots, \xi_n, \eta_1, \ldots, \eta_n]}{(\xi_1 + i\eta_1, \ldots, \xi_n + i\eta_n)}$$

considered as a $\mathcal{T}_{2n}$-module via the map

$$\alpha\colon \mathbb{C}[\xi_1, \ldots, \xi_n, \eta_1, \ldots, \eta_n] \to \mathbb{C}[\xi_1, \ldots, \xi_n]$$

defined by

$$p(\xi_1, \ldots, \xi_n, \eta_1, \ldots, \eta_n) \to p(\xi_1, \ldots, \xi_n, i\xi_1, \ldots, i\xi_n)$$

for $p \in \mathcal{T}_{2n}$.

It follows that the simple complex associated to the tensor product complex of (1*) and (2*) is a Hilbert resolution ([4] theorem 3) of the module M considered as a $\mathcal{T}_{2n}$-module via the mapping α.

If we introduce the complex coordinates

$$\zeta_1 = \xi_1 - i\eta_1, \ldots, \zeta_n = \xi_n - i\eta_n$$

$$\bar{\zeta}_1 = \xi_1 + i\eta_1, \ldots, \bar{\zeta}_n = \xi_n + i\eta_n$$

and set

$$\mathcal{T} = \mathbb{C}[\zeta_1, \dots, \zeta_n] = \frac{\mathbb{C}[\zeta_1, \dots, \zeta_n, \bar{\zeta}_1, \dots, \bar{\zeta}_n]}{(\zeta_1, \dots, \zeta_n)},$$

we can consider M as a $\mathcal{T}$-module (and thus as a $\mathbb{C}[\zeta_1, \dots, \zeta_n, \bar{\zeta}_1, \dots, \bar{\zeta}_n]$ module) given by the « economic » presentation

$$0 \leftarrow M \leftarrow \mathcal{T}^{p_0} \xleftarrow{{}^tA_0(\zeta)} \mathcal{T}^{p_1}.$$

b) Let M be a $\mathcal{T}$-module of finite type. For $m \in M$ we set

$$\operatorname{Ann}(m) = \{p \in \mathcal{T} | pm = 0\}.$$

This is an ideal of $\mathcal{T}$.

The set of $\operatorname{Ann}(m)$ which are prime ideals is called the set of the associated ideals to M and is denoted by $\operatorname{Ass}(M)$. It is known to be a finite set of prime ideals ([5] Algèbre Commutative ch. 4 § 1 n. 4 th. 2).

We also define

$$\operatorname{Ann}(M) = \bigcap_{m \in M} \operatorname{Ann}(m)$$

$$\operatorname{Supp}(M) = \{\mathfrak{p} \subset \mathcal{T} \text{ prime ideal with } \mathfrak{p} \supset \operatorname{Ann}(M)\}$$

([5] Algèbre Commutative ch. 2 § 4 n. 4 proposition 17).

It is also known ([5] Algèbre Commutative ch. 4 § 1 n. 4 th. 2) that $\operatorname{Ass}(M)$ and $\operatorname{Supp}(M)$ have the same minimal elements.

We do have therefore

$$\begin{aligned}\sqrt{\operatorname{Ann}(M)} &= \bigcap_{\substack{\mathfrak{p} \supset \operatorname{Ann}(M) \\ \mathfrak{p} \text{ prime}}} \mathfrak{p} \\ &= \bigcap_{\mathfrak{p} \in \operatorname{Supp}(M)} \mathfrak{p} \\ &= \bigcap_{\substack{\mathfrak{p} \in \operatorname{Supp} M \\ \mathfrak{p} \text{ minimal}}} \mathfrak{p} \\ &= \bigcap_{\mathfrak{p} \in \operatorname{Ass}(M)} \mathfrak{p}.\end{aligned}$$

We say that the module $M \neq 0$ is *coprimary* if $\forall m \in M$, $m \neq 0$,

$$\operatorname{Ann}(m) \subset \sqrt{\operatorname{Ann}(M)}.$$

In this case $\sqrt{\operatorname{Ann}(M)}$ reduces to a prime ideal $\mathfrak{p}$ and at the same time we must have $\operatorname{Ass}(M) = \{\mathfrak{p}\}$. We also say that *$M$ is $\mathfrak{p}$-coprimary.*

Given a $\mathcal{S}$-module M of finite type and a submodule $N \subset M$ we will say that N is *$\mathfrak{p}$-primary* (in M) if M/N is $\mathfrak{p}$-coprimary.

Given a $\mathcal{S}$-module M of finite type and a submodule N, for any prime ideal $\mathfrak{p} \in \operatorname{Ass}(M/N)$ we can find a $\mathfrak{p}$-primary submodule $Q(\mathfrak{p})$ of M such that

$$N = \bigcap_{\mathfrak{p}\in\operatorname{Ass}(M/N)} Q(\mathfrak{p}) .$$

We have moreover if $\operatorname{Ass}(M/N) = \mathfrak{p}_1 \cup \ldots \cup \mathfrak{p}_k$

$$(\alpha) \qquad \bigcap_{j\neq i} Q(\mathfrak{p}_j) \not\subset Q(\mathfrak{p}_i) \qquad \forall i \in \{1, \ldots, k\}$$

(β) $\operatorname{Ass}(M/Q(\mathfrak{p}_i)) = \mathfrak{p}_i$ and the $\mathfrak{p}_i$'s are two by two distinct.

If a module $N \subset M$ is written as $N = \bigcap_{i=1}^{k} Q(\mathfrak{p}_i)$ with $Q(\mathfrak{p}_i)$ $\mathfrak{p}_i$-primary and satisfying conditions (α) and (β) we say that we have written a reduced primary decomposition of N.

In every reduced primary decomposition of the submodule N of M the prime ideals $\mathfrak{p}_i$ are uniquely determined and are the elements of $\operatorname{Ass}(M/N)$; moreover if $\mathfrak{p}_i$ is a minimal element of $\operatorname{Ass}(M/N)$ then $Q(\mathfrak{p}_i)$ is also uniquely determined.

c) Let us go back to the notations of point *a*) with the presentation of the module M of finite type

$$0 \leftarrow M \leftarrow \mathcal{S}^{p_0} \xleftarrow{{}^tA_0(\zeta)} \mathcal{S}^{p_1}$$

and let us consider the submodule N of $\mathcal{S}^{p_0}$ given by

$$N = \operatorname{Im} {}^tA_0(\zeta) .$$

Let $\operatorname{Ass}(M) = \mathfrak{p}_1 \cup \ldots \cup \mathfrak{p}_k$. For every prime ideal $\mathfrak{p}_i$ let us take a primary submodule $Q(\mathfrak{p}_i)$ of $\mathcal{S}^{p_0}$ so that

$$N = \bigcap_{\mathfrak{p}_i\in\operatorname{Ass}(M)} Q(\mathfrak{p}_i)$$

is a reduced primary decomposition of N.

Set $M_i = \mathcal{S}^{p_0}/Q(\mathfrak{p}_i)$ so that $\operatorname{Ass}(M_i) = \mathfrak{p}_i$ and let us consider for each $i = 1, \ldots, k$ a presentation of the module M_i

$$0 \leftarrow M_i \leftarrow \mathcal{S}^{p_0} \xleftarrow{{}^tA_i(\zeta)} \mathcal{S}^{q_i}, \qquad 1 \leqslant i \leqslant k ,$$

where ${}^tA_i(\zeta)$ is a matrix with entires polynomials in ζ with p_0 rows and q_i columns.

As

$$\operatorname{Im} {}^tA_i(\zeta) = Q(\mathfrak{p}_i) \supset N = \operatorname{Im} {}^tA_0(\zeta)$$

we can find matrices ${}^t\lambda_i(\zeta)$ with polynomial entries such that we get commutative diagrams

$$\begin{array}{ccc} \mathfrak{P}^{p_1} & \xrightarrow{\;{}^tA_0(\zeta)\;} & \\ {}^t\lambda_i(\zeta)\Big\downarrow & & \mathfrak{P}^{p_0}\,, \qquad 1 \leqslant i \leqslant k\,, \\ \mathfrak{P}^{q_i} & \xrightarrow{\;{}^tA_i(\zeta)\;} & \end{array}$$

i.e. ${}^tA_0(\zeta) = {}^tA_i(\zeta)\,{}^t\lambda_i(\zeta)$ and thus the identity of differential operators

$$A_0(D) = \lambda_i(D)\,A_i(D)\,, \qquad 1 \leqslant i \leqslant k\,.$$

We denote by V_i the algebraic subvariety of $\mathbb{C}^n$ associated to the prime ideal $\mathfrak{p}_i$:

$$V_i = \{\xi \in \mathbb{C}^n \mid p(\xi) = 0 \;\; \forall p \in \mathfrak{p}_i\}\,, \qquad 1 \leqslant i \leqslant k\,,$$

so that $V = \bigcup_{i=1}^{k} V_i$ is the algebraic variety called the characteristic variety of the differential operator $A_0(D)$.

d) Let P and Q denote two polynomials in n variables. We recall the following, useful, identity

$$P(D_z)\big(\exp(\langle \xi, z\rangle) Q(z)\big) = Q(D_\xi)\big(\exp(\langle \xi, z\rangle) P(\xi)\big)$$

where $z = (z_1, \ldots, z_n)$, $\xi = (\xi_1, \ldots, \xi_n)$ are complex coordinates in $\mathbb{C}^n$ and where

$$\langle \xi, z\rangle = \sum_{i=1}^{n} \xi_i z_i\,.$$

Indeed, using Leibnitz formula, right and left hand side of the above equality are expressed by

$$\sum_\alpha \frac{1}{\alpha!} Q^{(\alpha)}(z) P^{(\alpha)}(\xi) \exp[\langle \xi, z\rangle]\,.$$

In particular if P is replaced by a (p, q)-matrix with polynomial entries and Q is replaced by a (q, r) matrix also with polynomial entries we have

$$P(D_z)\big(\exp(\langle\xi, z\rangle)Q(z)\big) = {}^t\{{}^tQ(D_\xi)\big(\exp(\langle\xi, z\rangle)\,{}^tP(\xi)\big)\}\ .$$

e) Let us go back to the primary decomposition $N = \bigcap_{i=1}^{k} Q(\mathfrak{p}_i)$ of the submodule $N = \operatorname{Im}{}^tA_0(\xi)$ of $\mathcal{P}^{p_0}$.

By a *theorem of Palamodov* [cf. 17].

For every i, $1 \leqslant i \leqslant k$, we can construct a matrix $\mathfrak{L}_i(z, \xi)$ of type $(*, p_0)$ with entries polynomials in z and ξ with the following property:

The necessary and sufficient condition for $X = \begin{pmatrix} X_1(\xi) \\ \vdots \\ X_{p_0}(\xi) \end{pmatrix} \in \mathcal{P}^{p_0}$ *to be contained in the primary submodule $Q(\mathfrak{p}_i)$ is that*

$$\mathfrak{L}_i(\xi, D_\xi)X(\xi) = 0$$

for every $\xi \in V_i$ (the algebraic variety of the prime ideal $\mathfrak{p}_i$). In particular we will have

$$\mathfrak{L}_i(\xi, D_\xi)\big(\exp(\langle\xi, z\rangle)\,{}^tA_i(\xi)\big) = 0 \qquad \forall\xi \in V_i$$

because $\exp(\langle\xi, z\rangle) = \sum_{l=0}^{\infty} \langle\xi, z\rangle^l/l!$ can be uniformly approximated by polynomials.

By virtue of the formula quoted above we then obtain that for $\forall z \in \mathbb{C}^n$ and $\forall\xi \in V_i$ we have

$$A_i(D_z)\big(\exp(\langle\xi, z\rangle)\,{}^t\mathfrak{L}_i(\xi, z)\big) = 0\ .$$

In particular as $A_0(D_z) = \lambda_i(D_z)A_i(D_z)$ we get $\forall z \in \mathbb{C}^n$, $\forall\xi \in V_i$

$$A_0(D_z)\big(\exp(\langle\xi, z\rangle\,{}^t\mathfrak{L}_i(\xi, z)\big) = 0\ .$$

This means that any column $u(z)$ of the $(p_0, *)$ matrix

$$\exp(\langle\xi, z\rangle)\,{}^t\mathfrak{L}_i(\xi, z)$$

for any fixed $\xi \in V_i$ as a function of $z \in \mathbb{C}^n$ is an exponential polynomial solution of the equation $A_i(D_z)u = 0$ and thus also of $A_0(D_z)u = 0$.

The operators $\mathfrak{L}_i(\xi, D_\xi)$ are sometimes called the *noetherian operators* associated with the operator $A_0(D_z)$.

f) We now consider the space $\Gamma(\mathbb{C}^n, \mathcal{O}_{A_0})$ as a subspace of the space $\Gamma(\mathbb{C}^n, \mathcal{E}^{p_0})$ of C^∞ functions defined in $\mathbb{C}^n$ with values in $\mathbb{C}^{p_0}$. We can endow the first of these spaces with the topology of uniform convergence on compact sets and the second with its Schwartz topology of uniform convergence on compact sets of the functions and all their partial derivatives. The inclusion map

$$\Gamma(\mathbb{C}^n, \mathcal{O}_{A_0}) \to \Gamma(\mathbb{C}^n, \mathcal{E}^{p_0})$$

is then continuous and the image of the first space is closed. It follows that the Schwartz topology induces on $\Gamma(\mathbb{C}^n, \mathcal{O}_{A_0})$ a topology equivalent with the original one (a fact that could also be established directly by means of Cauchy integral formula).

Set

$$\mathcal{O}'_{A_0}(\mathbb{C}^n) = \text{Hom cont}\big(\Gamma(\mathbb{C}^n, \mathcal{O}_{A_0}), \mathbb{C}\big)$$
$$\mathcal{E}'^{p_0}(\mathbb{C}^n) = \text{Hom cont}\big(\Gamma(\mathbb{C}^n, \mathcal{E}^{p_0}), \mathbb{C}\big)\,.$$

Given a continuous functional $\mu \in \mathcal{O}'_{A_0}(\mathbb{C}^n)$ this is defined on $\Gamma(\mathbb{C}^n, \mathcal{O}_{A_0})$ and it is continuous also when this space is considered as a subspace of $\Gamma(\mathbb{C}^n, \mathcal{E}^{p_0})$.

By the theorem of Hahn-Banach μ admits a continuous extension

$$\eta = \begin{pmatrix} \eta_1 \\ \vdots \\ \eta_{p_0} \end{pmatrix} \in \mathcal{E}'^{p_0}(\mathbb{C}^n)\,,$$

where for

$$\varphi = \begin{pmatrix} \varphi_1 \\ \vdots \\ \varphi_{p_0} \end{pmatrix} \in \Gamma(\mathbb{C}^n, \mathcal{E}^{p_0})$$

we have by definition

$$\langle \eta, \varphi \rangle = \sum_{i=1}^{p_0} \eta_i(\varphi_i)\,.$$

Once μ is extended with η we can consider the Laplace transform of η.

This by definition is a p_0-uple of entire holomorphic functions on $\mathbb{C}^n$

$$\tilde{\eta}(\xi) = \begin{pmatrix} \tilde{\eta}_1(\xi) \\ \vdots \\ \tilde{\eta}_{p_0}(\xi) \end{pmatrix} \quad \text{where}$$

$$\tilde{\eta}_j(\xi) = \eta_j\big(\exp(\langle \xi, z \rangle)\big)\,.$$

We remark the following

Let $\mathcal{L}_i(\xi, D_\xi)$ be the noetherian operator associated to the algebraic variety V_i $(i = 1, \ldots, k)$. Let $\tilde{\eta}(\xi)$ be the Laplace transform of an extension η to $\Gamma(\mathbf{C}^n, \mathcal{E}^{p_0})$ of a functional μ defined on $\Gamma(\mathbf{C}^n, \mathcal{O}_{A_0})$.

Then the value of

$$\mathcal{L}_i(\xi, D_\xi)\tilde{\eta}(\xi)|V_i$$

is independent from the choice of the extension η of μ.

Indeed we have, suppressing the index i and setting $\mathcal{L} = (\mathcal{L}_{\alpha\beta})$,

$$\begin{aligned}\sum_\beta \mathcal{L}_{\alpha\beta}(\xi, D_\xi)\tilde{\eta}_\beta(\xi) &= \sum_\beta \mathcal{L}_{\alpha\beta}(\xi, D_\xi)\eta_\beta[\exp(\langle \xi, z\rangle)] \\ &= \sum_\beta \eta_\beta[\mathcal{L}_{\alpha\beta}(\xi, D_\xi)\exp(\langle \xi, z\rangle)] \\ &= \sum_\beta \eta_\beta[\exp(\langle \xi, z\rangle)\mathcal{L}_{\alpha\beta}(\xi, z)] \\ &= \sum_\beta \mu_\beta[\exp(\langle \xi, z\rangle)\mathcal{L}_{\alpha\beta}(\xi, z)] \qquad \text{for } \xi \in V_i\end{aligned}$$

i.e. in matrix notations

$$\mathcal{L}_i(\xi, D_\xi)\,\tilde{\eta}(\xi) = \langle \mu, \exp(\langle \xi, z\rangle)\,{}^t\mathcal{L}_i(\xi, z)\rangle \qquad \text{for } \xi \in V_i$$

because of the remark that the columns of the matrix of type $(p_0, *)$ $\exp(\langle \xi, z\rangle)\,{}^t\mathcal{L}_i(\xi, z)$ for $\xi \in V_i$ are elements of $\Gamma(\mathbf{C}^n, \mathcal{O}_{A_0})$, and for

$$\mu = \begin{pmatrix}\mu_1\\ \vdots \\ \mu_{p_0}\end{pmatrix} \in \mathcal{O}'_{A_0}(\mathbf{C}^n) \qquad \text{and} \qquad u(z) = \begin{pmatrix}u_1(z)\\ \vdots \\ u_{p_0}(z)\end{pmatrix} \in \Gamma(\mathbf{C}^n, \mathcal{O}_{A_0})$$

we have

$$\langle \mu, u\rangle = \sum_{j=1}^{p_0} \mu_j[u_j]\,.$$

g) For $u \in \Gamma(\mathbf{C}^n, \mathcal{O}^{p_0})$ for K compact in $\mathbf{C}^n$ and $m \geqslant 0$ an integer we set

$$\|u\|_{K,m} = \sup_{z\in K} \sum_{|\alpha|\leqslant m} |D_z^\alpha u(z)|$$

where $|\cdot|$ denotes a norm on $\mathbf{C}^{p_0}$.

Also we set

$$H_K(\xi) = \sup_{z\in K} \operatorname{Re}\langle \xi, z\rangle\,.$$

We can now state a form of *Ehrenpreis fundamental principle* [cf. 9]. *Let* μ *be a continuous functional on* $\Gamma(\mathbb{C}^n, \mathcal{O}_{A_0})$ *and let* η *be a continuous extension* of μ to $\Gamma(\mathbb{C}^n, \mathcal{E}^{p_0})$ *and let* $\tilde{\eta} = \tilde{\eta}(\xi)$ *denote the Laplace transform of* η. *Let* K *be a compact convex set in* $\mathbb{C}^n$

a) Assume that we have estimates of the form (with constants $c > 0$)

$$|\mathcal{L}_i(\xi, D_\xi)\,\tilde{\eta}(\xi)| \leqslant c(1 + |\xi|)^m \exp(H_K(\xi)) \qquad \forall \xi \in V_i$$

and for $i = 1, \dots, k$.

Then we have for any $u \in \Gamma(\mathbb{C}^n, \mathcal{O}_{A_0})$ *estimates of the form*

$$|\mu(u)| \leqslant cc_0 \|u\|_{K,m+N}$$

where $c_0 = c_0(m, K, A_0)$ *and* $N = N(A_0)$.

(b) Assume that for any $u \in \Gamma(\mathbb{C}^n, \mathcal{O}_{A_0})$ *we have estimates of the form*

$$|\mu(u)| \leqslant c\|u\|_{K,m}$$

with a positive constant c.

Then we have also estimates of the form

$$|\mathcal{L}_i(\xi, D_\xi)\,\tilde{\eta}(\xi)| \leqslant cc_1(1 + |\xi|)^{m+N_1} \exp(H_K(\xi)) \qquad \forall \xi \in V_i$$

and for $i = 1, \dots, k$, *where* $c_1 = c_1(K, m, A_0)$ *and* $N_1 = N_1(A_0)$.

SECTION 9

First Transformation of the Carrier Condition (C)

We consider the $\bar{\partial}$-suspension to $\mathbb{C}^n$ of the Hilbert complex (1).

Let ω be an open convex set in $\mathbb{R}^n$. Then ω is analytically convex with respect to the Hilbert complex (1) *i.e.*

$$H^1(\omega, \mathcal{A}_{A_0}) = 0$$

if and only if the following carrier condition is satisfied

(*C*) *Given any convex compact set $K \subset \omega$ we can find $\delta > 0$ and a convex compact set K' with $K \subset K' \subset \omega$ such that given any convex compact set $K'' \subset \omega$, for any analytic functional $\mu \in \mathcal{O}'_{A_0}(\mathbb{C}^n)$ having the property*

$$\mu \text{ has a carrier in } K_\delta = \{z \in \mathbb{C}^n | \text{dist}\,(z, K) \leqslant \delta\}$$

$$\mu \text{ has a carrier in } K''$$

we have also that

$$\mu \text{ has a carrier in } K'.$$

In $\mathbb{C}^n$ we consider the euclidean norm and the euclidean distance. Given K compact in $\mathbb{C}^n$ we set for any $\varepsilon > 0$.

$$K_\varepsilon = \{z \in \mathbb{C}^n | \text{dist}\,(z, K) \leqslant \varepsilon\}.$$

We have the following equality

$$H_{K_\varepsilon}(\xi) = H_K(\xi) + \varepsilon|\xi|.$$

Indeed set for $\xi = \xi' + i\xi'' \in \mathbb{C}^n$, $\xi_{\mathbf{R}} = (\xi_1', \ldots, \xi_n', \xi_1'', \ldots, \xi_n'')$ so that denoting with a point the usual scalar product in $\mathbb{R}^{2n} = \mathbb{C}^n$ we have $\operatorname{Re}\langle \xi, z\rangle = \bar{\xi}_{\mathbf{R}} \cdot z_{\mathbf{R}}$.

Now let $H_K(\xi) = \bar{\xi}_{\mathbf{R}} \cdot z_{\mathbf{R}}^{(0)}$ with $z_{\mathbf{R}}^{(0)} \in K$. Then we have

$$H_{K_\varepsilon}(\xi) = \sup_{z \in K_\varepsilon} \bar{\xi}_{\mathbf{R}} \cdot z_{\mathbf{R}} \geqslant \bar{\xi}_{\mathbf{R}} \cdot \left(z_{\mathbf{R}}^{(0)} + \varepsilon \frac{\xi_{\mathbf{R}}}{|\xi|} \right) \geqslant H_K(\xi) + \varepsilon|\xi| \, .$$

Also let $H_{K_\varepsilon}(\xi) = \bar{\xi}_{\mathbf{R}} \cdot z_{\mathbf{R}}^{(1)}$ with $z_{\mathbf{R}}^{(1)} \in K_\varepsilon$. We can find $z_{\mathbf{R}}^{(0)} \in K$ and $\nu_{\mathbf{R}}$ with $|\nu_{\mathbf{R}}| = 1$ such that $z_{\mathbf{R}}^{(0)} + \varepsilon \nu_{\mathbf{R}} = z_{\mathbf{R}}^{(1)}$. Then

$$\begin{aligned} H_{K_\varepsilon}(\xi) &= \bar{\xi}_{\mathbf{R}} \cdot z_{\mathbf{R}}^{(1)} = \bar{\xi}_{\mathbf{R}} \cdot (z_{\mathbf{R}}^{(0)} + \varepsilon \nu_{\mathbf{R}}) \\ &\leqslant H_K(\xi) + \varepsilon|\xi| \quad \text{(by Schwarz inequality)} \, . \end{aligned}$$

This proves our contention.

We also remark that for every positive integer m and any $\varepsilon > 0$ there exists a constant $c(m, \varepsilon)$ such that for any $u \in \Gamma(K_\varepsilon, \mathcal{O}^{p_0})$ we have an estimate

$$\|u\|_{K,m} \leqslant c(m, \varepsilon) \|u\|_{K_\varepsilon}$$

(where $\|u\|_{K_\varepsilon}$ stands for $\|u\|_{K_\varepsilon, 0}$).

Finally we remark that for any $N > 0$, any $\varepsilon > 0$ there exists a constant $c(N, \varepsilon) > 0$ such that

$$(1 + |\xi|)^N \leqslant c(N, \varepsilon) \exp(\varepsilon|\xi|) \, .$$

Let now $\mu \in \mathcal{O}'_{A_0}(\mathbb{C}^n)$ and let η be any continuous extension of μ from $\Gamma(\mathbb{C}^n, \mathcal{O}_{A_0})$ to $\Gamma(\mathbb{C}^n, \mathcal{E}^{p_0})$. Let $\tilde{\eta}(\xi) = \begin{pmatrix} \tilde{\eta}_1(\xi) \\ \vdots \\ \tilde{\eta}_{p_0}(\xi) \end{pmatrix}$ denote the Laplace transform of η. We deduce from the previous remarks the following

PROPOSITION 7. *The necessary and sufficient condition for the «A_0-analytic functional» μ to have a carrier in a given compact convex K is that for the Laplace transform we have an estimate of the form:*

for every $\varepsilon > 0$ there exists a constant $c(\varepsilon) > 0$ such that for $1 \leqslant i \leqslant k$ we have

$$|\mathcal{L}_i(\xi, D_\xi)\tilde{\eta}(\xi)| \leqslant c(\varepsilon) \exp\left(H_K(\xi) + \varepsilon|\xi|\right)$$

for every $\xi \in V_i$:

Using this remark we deduce the following equivalent form of the carrier condition for an open convex set ω in $\mathbb{R}^n$

(*C*) *Given any convex compact set* $K \subset \omega$ *we can find* $\delta > 0$ *and a convex compact set* K' *with* $K \subset K' \subset \omega$ *such that*

given a convex compact set $K'' \subset \omega$, *for any « A_0-analytic functional »* $\mu \in \mathcal{O}'_{A_0}(\mathbb{C}^n)$, *for any continuous extension* η *to* $\Gamma(\mathbb{C}^n, \mathcal{E}^{p_0})$ *we have for the Laplace transform* $\tilde{\eta}$ *of* η *the following condition:*

Assume that for every $\varepsilon > 0$ *we can find a constant* $c_\varepsilon > 0$ *such that for* $1 \leqslant i \leqslant k$ *we have*

$$|\mathfrak{L}_i(\xi, D_\xi)\tilde{\eta}(\xi)| \leqslant c_\varepsilon \exp\big(H_K(\xi) + (\delta + \varepsilon)|\xi|\big) \qquad \forall \xi \in V_i$$

$$|\mathfrak{L}_i(\xi, D_\xi)\tilde{\eta}(\xi)| \leqslant c_\varepsilon \exp\big(H_{K''}(\xi) + \varepsilon|\xi|\big) \qquad \forall \xi \in V_i$$

then we have also that for every $\varepsilon > 0$ *we can find a constant* $c'_\varepsilon > 0$ *such that, for* $1 \leqslant i \leqslant k$, *we have*

$$|\mathfrak{L}_i(\xi, D_\xi)\tilde{\eta}(\xi)| \leqslant c'_\varepsilon \exp\big(H_{K'}(\xi) + \varepsilon|\xi|\big) \qquad \forall \xi \in V_i\,.$$

Now we remark that $\tilde{\eta}(\xi)$, and consequently $\mathfrak{L}_i(\xi, D_\xi)\tilde{\eta}(\xi)$, is an entire holomorphic function.

We conclude then with the following

PROPOSITION 8. *Let* ω *be open and convex in* $\mathbb{R}^n$. *Assume that for any compact set* $K \subset \omega$ *we can find* $\delta > 0$ *and a convex compact set* K' *with* $K \subset K' \subset \omega$ *such that*

given any convex compact set $K'' \subset \omega$, *any entire holomorphic function* $F(\xi)$ *in* $\mathbb{C}^n$ *with the property*

$\forall \varepsilon > 0$ *there exists a constant* a_ε *such that, for* $1 \leqslant i \leqslant k$,

$$\log|F(\xi)| \leqslant a_\varepsilon + H_K(\xi) + (\delta + \varepsilon)|\xi| \qquad \forall \xi \in V_i$$

$$\log|F(\xi)| \leqslant a_\varepsilon + H_{K''}(\xi) + \varepsilon|\xi| \qquad \forall \xi \in V_i$$

has also the property

$\forall \varepsilon > 0$ *there exists a constant* a'_ε *such that, for* $1 \leqslant i \leqslant k$,

$$\log|F(\xi)| \leqslant a'_\varepsilon + H_{K'}(\xi) + \varepsilon|\xi| \qquad \forall \xi \in V_i\,.$$

Then ω *is analytically convex i.e.*

$$H^1(\omega, \mathcal{A}_{A_0}) = 0 \,.$$

We recall that the varieties V_i are the algebraic varieties associated to the different primes $\mathfrak{p}_i$ for a primary decomposition in $\mathcal{P}^{p_0}$ of the module

$$\operatorname{Im}\left(\mathcal{P}^{p_1} \xrightarrow{{}^tA_0(\xi)} \mathcal{P}^{p_0}\right)$$

$1 \leqslant i \leqslant k$.

We also use the convention, now and in the sequel, that $\log 0 = -\infty$.

SECTION 10

The Gauge Function of Entire Functions of Exponential Type.

a) Let $F(z)$ be an entire function defined on $\mathbb{C}^n$. We will say that $F(z)$ is of exponential type if one can find positive constants $c_1 > 0$, $c > 0$ such that

$$|F(z)| \leqslant c_1 \exp(c|z|) \qquad \forall z \in \mathbb{C}^n$$

(by $|\cdot|$ we denote the euclidean norm on $\mathbb{C}^n$).

Let V be an irreducible algebraic subvariety of $\mathbb{C}^n$. A function $F(z)$ defined and holomorphic on V will be called of exponential type if for some positive constants $c_1 > 0$, $c > 0$, we have

$$|F(z)| \leqslant c_1 \exp(c|z|) \qquad \forall z \in V .$$

If $V = \mathbb{C}^n$ we recover the previous definition. Moreover it is worth noticing the following theorem which is an essential tool in the proof of Ehrenpreis fundamental principle

EXTENSION THEOREM. *Let V be an irreducible algebraic variety of $\mathbb{C}^n$ and let F be a holomorphic function of exponential type defined on V so that*

$$|F(z)| \leqslant c_1 \exp(c|z|) \qquad \forall z \in V$$

with constants $c_1 > 0$, $c > 0$.

There exist an integer $N > 0$ and a new constant $c_2 > 0$ such that we can find an entire function $G(z)$ on $\mathbb{C}^n$ with the following properties:

i) $G(z) = F(z)$ $\forall z \in V$

ii) $|G(z)| \leqslant c_2(1 + |z|)^N \exp(c|z|)$ $\forall z \in \mathbb{C}^n$.

Let μ be an analytic functional on $\mathbb{C}^n$, i.e.

$$\mu \in \text{Hom cont}\,(\Gamma(\mathbb{C}^n, \mathcal{O}), \mathbb{C})\,.$$

By the continuity of μ there exists a positive constant c_1 and a compact set K such that

$$|\mu(f)| \leqslant c_1 \|f\|_K \qquad \forall f \in \Gamma(\mathbb{C}^n, \mathcal{O})\,.$$

In particular for the Laplace transform $\tilde{\mu}(\xi) = \mu(\exp(\langle \xi, z \rangle))$ we have an estimate

$$|\tilde{\mu}(\xi)| \leqslant c_1 \exp(H_K(\xi)) \qquad \forall \xi \in \mathbb{C}^n$$

so that for some positive constant $c > 0$ we have

$$|\tilde{\mu}(\xi)| \leqslant c_1 \exp(c|\xi|)\,.$$

This shows that the Laplace transform of an analytic functional is an entire function on $\mathbb{C}^n$ of exponential type.

b) Let V be an irreducible algebraic subvariety of $\mathbb{C}^n$ of $\dim_{\mathbb{C}} V \geqslant 1$ and let $\mathfrak{b}$ denote the (prime) ideal of polynomials vanishing on V. Let $\mathfrak{a}$ be the asymptotic ideal of the ideal $\mathfrak{b}$ i.e. the homogeneous ideal of the principal parts of the polynomials of $\mathfrak{b}$. Let W denote the asymptotic variety of V i.e. the variety of common zeros in $\mathbb{C}^n$ of the elements of $\mathfrak{a}$. Then W is a cone with vertex at the origin in $\mathbb{C}^n$ and each irreducible component of W has the same dimension than the dimension of V. In other words W is a purely dimensional cone with vertex at the origin and of dimension equal to $\dim_{\mathbb{C}} V$.

Let $F = F(z)$ be a holomorphic function of exponential type defined on V i.e.

$$|F(z)| \leqslant c_1 \exp(c|z|) \qquad \forall z \in V\,.$$

For $\Theta \in W \cap \{z \in \mathbb{C}^n \mid |z| = 1\}$ we define

$$\gamma(\Theta) = \max \lim_{\substack{z \in V \\ |z| \to \infty \\ \frac{z}{|z|} \to \Theta}} \frac{\log |F(z)|}{|z|}\,.$$

For $|z| \geqslant 1$ we have

$$\frac{\log |F(z)|}{|z|} \leqslant \frac{\log c_1}{|z|} + c \leqslant \log c_1 + c$$

so that the left hand side is uniformly bounded from above and then $\gamma(\Theta)$ is well defined on $W \cap \{z \in \mathbb{C}^n | \, |z| = 1\}$.

Now we extend $\gamma(\Theta)$ to the whole variety W as a homogeneous function of degree one by defining for every $\Theta \in W - \{0\}$

$$g(\Theta) = |\Theta| \, \gamma\left(\frac{\Theta}{|\Theta|}\right)$$

and

$$g(0) = \max_{\substack{\Theta \in W - \{0\} \\ \Theta \to 0}} \lim g(\Theta) .$$

Unless $\gamma(\Theta) \equiv -\infty$ i.e. unless $F(z) \equiv 0$ on V we have

$$g(0) = 0$$

(otherwise $g(\Theta) \equiv -\infty$ on W).

The function g defined in this way on the asymptotic variety W is called the *gauge function* on W associated to the holomorphic function F of exponential type defined on V.

PROPOSITION 9. *Let F be a holomorphic function defined on V and of exponential type. Let g be the associated gauge function of F defined on the asymptotic variety W of V. Let K be a compact set of $\mathbb{C}^n$.*

The following statements are equivalent

i) $\forall \Theta \in W$ *we have the estimate*

$$g(\Theta) \leqslant H_K(\Theta)$$

ii) $\forall \varepsilon > 0$ *there exists a constant* $c_\varepsilon > 0$ *such that*

$$|F(z)| \leqslant c_\varepsilon \exp\left(H_K(z) + \varepsilon|z|\right) \qquad \forall z \in V .$$

PROOF. ii) $\Rightarrow$ i). For every $\Theta \in W - \{0\}$ we can find a sequence $\{z_\nu\}_{\nu \in \mathbf{N}} \subset V - \{0\}$ such that

$$|z_\nu| \to \infty, \qquad \frac{z_\nu}{|z_\nu|} \to \frac{\Theta}{|\Theta|} \qquad \text{and} \qquad |\Theta| \frac{\log |F(z_\nu)|}{|z_\nu|} \to g(\Theta) .$$

By assumption we have

$$|\Theta| \frac{\log |F(z)|}{|z|} \leqslant |\Theta| \frac{\log c_\varepsilon}{|z|} + |\Theta| \, H_K\left(\frac{z}{|z|}\right) + \varepsilon |\Theta| .$$

Passing to the limit for $z = z_\nu$ and $\nu \to \infty$ we get

$$g(\Theta) \leqslant H_K(\Theta) + \varepsilon|\Theta| .$$

Since ε is arbitrary, this implies

$$g(\Theta) \leqslant H_K(\Theta) ,$$

a relation which is also true when $\Theta = 0$.

Before proving the opposite implication we establish the following

LEMMA. *For any* $\varrho > 0$ *we can find* $M = M(\varrho) > 0$ *such that* $\forall z \in V$ *with* $|z| > M$ *we have* $\operatorname{dist}(z, W) < \varrho|z|$.

If this statement is false there must exist a $\varrho > 0$ and a sequence $\{z_\nu\}_{\nu\in\mathbf{N}} \subset V$ with

$$|z_\nu| > \nu \quad \forall\nu \qquad \text{and} \qquad \operatorname{dist}(z_\nu, W) \geqslant \varrho|z_\nu| .$$

Passing to a subsequence, that we still will denote by $\{z_\nu\}$, we may assume that for $\nu \to \infty$,

$$\frac{z_\nu}{|z_\nu|} \to \Theta_0$$

for some $\Theta_0 \in \mathbf{C}^n$ with $|\Theta_0| = 1$. Since $|z_\nu| \to \infty$ we must have $\Theta_0 \in W$.

On the other hand since W is a cone $\operatorname{dist}(z, W)$ as a function of z is positively homogeneous of degree one so that we have

$$\operatorname{dist}\left(\frac{z_\nu}{|z_\nu|}, W\right) \geqslant \varrho .$$

Taking the limit for $\nu \to \infty$ we get $\operatorname{dist}(\Theta_0, W) \geqslant \varrho$. This is absurd since $\Theta_0 \in W$.

i) $\Rightarrow$ ii). Let $\varepsilon > 0$ be given and let

$$c = \sup_{z\in K} |z| .$$

For every $\Theta \in W \cap \{z \in \mathbf{C}^n \mid |z| = 1\}$ we can find $M = M(\Theta) > 0$ and $\sigma = \sigma(\Theta) > 0$ with $0 < \sigma < \min(\varepsilon/2c, 1)$ such that for $z \in V$, $|z| > M$,

$\left|z/|z| - \Theta\right| < \sigma$ we have

$$\frac{\log |F(z)|}{|z|} < g(\Theta) + \frac{\varepsilon}{2}.$$

This by the very definition of $g(\Theta) = \gamma(\Theta)$ as a « max lim ».

For $\alpha \in \mathbb{C}^n$, $r > 0$ we denote by $B(\alpha, r)$ the open ball with center α and radius r. By the compactness of $W \cap \{z \in \mathbb{C}^n \,\big|\, |z| = 1\}$ we can find $\Theta_1, \ldots, \Theta_s \in$ $\in W \cap \{z \in \mathbb{C}^n \,\big|\, |z| = 1\}$ such that

$$W \cap \{z \in \mathbb{C}^n \,|\, |z| = 1\} \subset \bigcup_1^s B\left(\Theta_i, \frac{\sigma(\Theta_i)}{2}\right).$$

Choose $M > \sup\limits_{1 \leqslant i \leqslant s} M(\Theta_i)$ and $\sigma = \frac{1}{4} \min\limits_{1 \leqslant i \leqslant s} \sigma(\Theta_i)$. If M is large enough, by the lemma, we may assume that

$$\operatorname{dist}(z, W) < \sigma |z| \qquad \text{for } \forall z \in V \text{ with } |z| > M.$$

Let z be a point of V with $|z| > M$. By the above statement, and as $\sigma < 1$, we can find $\Theta \in W - \{0\}$ with

$$\left|\frac{z}{|z|} - \Theta\right| < \sigma.$$

But then for $\eta = \Theta/|\Theta|$ we have

$$\begin{aligned} \left|\frac{z}{|z|} - \eta\right| &\leqslant \left|\frac{z}{|z|} - \Theta\right| + \big||\Theta| - 1\big| \\ &< \sigma + \left|\left|\frac{z}{|z|}\right| - |\Theta|\right| \\ &\leqslant \sigma + \left|\frac{z}{|z|} - \Theta\right| \\ &\leqslant 2\sigma. \end{aligned}$$

Since $|\eta| = 1$ and $\eta \in W$, for some i, $1 \leqslant i \leqslant s$, we must have $\eta \in B(\Theta_i, \sigma(\Theta_i)/2)$. This implies that

$$\left|\frac{z}{|z|} - \Theta_i\right| \leqslant \left|\frac{z}{|z|} - \eta\right| + |\eta - \Theta_i| < 2\sigma + \frac{\sigma(\Theta_i)}{2} \leqslant \sigma(\Theta_i).$$

By the remark at the beginning we thus have for that choice of z

$$\begin{aligned}\frac{\log|F(z)|}{|z|} &< g(\Theta_i) + \varepsilon/2 \\ &\leqslant H_K(\Theta_i) + \varepsilon/2 \\ &\leqslant H_K\left(\frac{z}{|z|}\right) + c\left|\frac{z}{|z|} - \Theta_i\right| + \varepsilon/2 \\ &\leqslant H_K\left(\frac{z}{|z|}\right) + \varepsilon .\end{aligned}$$

Therefore for any $z \in V$ with $|z| > M$ we have

$$(*) \qquad |F(z)| < \exp\left(H_K(z) + \varepsilon|z|\right) .$$

Now we remark that $C = V \cap \{z \in \mathbf{C}^n \mid |z| \leqslant M\}$ is compact so that

$$\delta = \inf_{z \in C} \exp\left(H_K(z) + \varepsilon|z|\right) > 0 .$$

Let

$$\sup_{z \in C} |F(z)| = \tau$$

and set $C_\varepsilon = 1 + \tau/\delta$. Then we have for any $z \in V$

$$|F(z)| \leqslant C_\varepsilon \exp\left(H_K(z) + \varepsilon|z|\right) .$$

Indeed for $z \in C$ the right hand side is greater than $\delta + \tau > \tau$ and for $z \in V - C$ the inequality follows from $(*)$ as $C_\varepsilon > 1$.

PROPOSITION 10. *Let g be the gauge function defined on the asymptotic variety W of the algebraic variety V associated to a holomorphic function F defined on V and of exponential type. For any $\Theta \in W$ we have*

$$g(\Theta) = \max \lim_{\substack{y \to \Theta \\ y \in W}} g(y) .$$

We will establish first some lemmas.

LEMMA 6. *The gauge function f is upper semicontinuous on W i.e. $\forall w \in W$ we have*

$$g(w) \geqslant \max \lim_{\substack{y \to w \\ y \in W}} g(y) .$$

PROOF. Let $w \neq 0$; we can find a sequence $\{y^{(n)}\} \subset W$ with $y^{(n)} \to w$ such that

$$\lim_{n\to\infty} g(y^{(n)}) = \max\lim_{y\to w} g(y) = \alpha .$$

For every n we can find $\xi^{(n)} \in V$, with

$$|\xi^{(n)}| > n , \qquad \left| \frac{\xi^{(n)}}{|\xi^{(n)}|} - \frac{y^{(n)}}{|y^{(n)}|} \right| < \frac{1}{n} ,$$

such that

$$(\star) \qquad \left| g(y^{(n)}) - |y^{(n)}| \frac{\log |F(\xi^{(n)})|}{|\xi^{(n)}|} \right| < \frac{1}{n} .$$

We have therefore, for $n \to \infty$, that

$$|\xi^{(n)}| \to \infty , \qquad \frac{\xi^{(n)}}{|\xi^{(n)}|} \to \frac{w}{|w|} .$$

We may suppose that $y^{(n)} \neq 0$ for every n. We have

$$\begin{aligned} g(w) &\geq \max\lim_{n\to\infty} |w| \frac{\log |F(\xi^{(n)})|}{|\xi^{(n)}|} \\ &= \max\lim_{n\to\infty} \left[\frac{|w|}{|y^{(n)}|} \left\{ |y^{(n)}| \frac{\log |F(\xi^{(n)})|}{|\xi^{(n)}|} - g(y^{(n)}) \right\} + \frac{|w|}{|y^{(n)}|} g(y^{(n)}) \right] = \alpha \end{aligned}$$

because, for $n \to \infty$, $|w|/|y^{(n)}| \to 1$ and because of the inequality $(*)$. If $w = 0$ we have $g(0) = \max\lim_{\substack{\Theta\in W-\{0\} \\ \Theta\to 0}} g(\Theta)$ by definition.

LEMMA 7. *Let $P(x, t) = P(x_1, \ldots, x_k, t)$ be a polynomial in the variables $x_1, \ldots, x_k, t$ with complex coefficients.*

We assume that the principal part of P is monic with respect to the variable t and of degree m.

We can find a constant ε with $0 < \varepsilon < 1$ (depending only on m, k, and the absolute value of the coefficients of P) with the following property:

for every $r > 0$ and for every point $(x^0, t^0) \in \mathbb{C}^{k+1}$ we can find a ϱ with $r/2 < \varrho < r$ such that

for every x with $|x - x^0| \leq \varepsilon \min(r, r^m)(1 + |(x^0, t^0)|)$

for every t with $|t - t^0| = \varrho(1 + |(x^0, t^0)|)$

we have

$$|P(x, t)| \geqslant \frac{1}{2}\left(\frac{r}{4(m+1)}\right)^m (1 + |(x^0, t^0)|)^m .$$

PROOF. α) Let $r > 0$ be given and let $(x_0, t_0) \in \mathbf{C}^{k+1}$ be fixed. We consider on the t plane the circular ring

$$\frac{r}{2}(1 + |(x^0, t^0)|) < |t - t_0| < r(1 + |(x^0, t^0)|) .$$

We divide this circular ring into $m+1$ concentric circular rings of equal width and let $R_0, R_1, \ldots, R_m$ denote their respective interiors so that

$$R_j : \frac{r}{2}(1 + |(x^0, t^0)|)\frac{j+m+1}{m+1} < |t - t_0| < \frac{r}{2}(1 + |(x^0, t^0)|)\frac{j+m+2}{m+1}$$

for $j = 0, 1, \ldots, m$, the width of each ring being

$$\frac{2(m+1)}{r}(1 + |(x^0, t^0)|) .$$

The equation in t $P(x^0, t) = 0$ has at most m distinct roots so that at least one R_{j0} of those rings does not contain any one of them.

Set

$$\varrho = \frac{r}{2}\,\frac{j_0 + m + 1}{m+1} + \frac{r}{4(m+1)}$$

so that $|t - t_0| = \varrho(1 + |(x^0, t^0)|)$ is a circle that divides R_{j0} into two circular rings of equal width. We have $r/2 < \varrho < r$.

For $|t - t^0| = \varrho(1 + |(x_0, t_0)|)$ and for any root τ of the equation $P(x^0, \tau) = 0$ we must have

$$|t - \tau| \geqslant \frac{r}{4(m+1)}(1 + |(x^0, t^0)|) .$$

Set $P(x^0, t) = (t - \tau_1) \ldots (t - \tau_m)$. From the above inequality we deduce that for every t with $|t - t_0| = \varrho(1 + |(x^0, t^0)|)$ we must have

$$|P(x^0, t)| \geqslant \left(\frac{r}{4(m+1)}\right)^m (1 + |(x^0, t^0)|)^m .$$

β) We have

$$P(x, t) - P(x^0, t) = \sum_{\substack{1 \leqslant |\alpha| \leqslant m \\ 0 \leqslant \beta \leqslant m - |\alpha|}} \frac{1}{\alpha!}\,\frac{1}{\beta!}\,(D_x^\alpha D_t^\beta P)_{(x^0, t^0)}(x - x^0)^\alpha (t - t^0)^\beta .$$

We set $|\alpha| + \beta = m - h$ and note that

$$|(x-x^0)^\alpha| \leqslant |x-x^0|^{|\alpha|} \leqslant (|x-x^0| + |t-t^0|)^{|\alpha|}$$
$$|(t-t^0)^\beta| = |t-t^0|^\beta \leqslant (|x-x^0| + |t-t^0|)^\beta .$$

Denoting by c (now and in the sequel of the proof) a constant depending only on m, the number of variables and the absolute value of the coefficients of P, we get

$$\begin{aligned}|P(x,t) - P(x^0,t)| &\leqslant c|x-x^0| \sum_{h=0}^{m-1} (1 + |(x^0,t^0)|)^h (|x-x^0| + |t-t^0|)^{m-h-1} \\ &\leqslant c|x-x^0|(1 + |(x^0,t^0)| + |x-x^0| + |t-t^0|)^{m-1} .\end{aligned}$$

Choose now ε with $0<\varepsilon<1$ and x, t such that

$$|x-x^0| \leqslant \varepsilon \min(r, r^m)(1 + |(x^0,t^0)|) , \qquad |t-t^0| < r(1 + |(x^0,t^0)|) .$$

We have

$$\begin{aligned}|P(x,t) - P(x^0,t)| &\leqslant \varepsilon c (1 + |(x^0,t^0)|)^m \min(r, r^m)(1 + \varepsilon \min(r, r^m) + r)^{m-1} \\ &\leqslant \varepsilon c (1 + |(x^0,t^0)|)^m \min(r, r^m) \max(1, r^{m-1}) \\ &\leqslant \varepsilon c r^m (1 + |(x^0,t^0)|)^m .\end{aligned}$$

γ) If we choose $\varepsilon > 0$ so small that

$$\varepsilon c < \frac{1}{2}\left(\frac{1}{4(m+1)}\right)^m$$

we obtain from the estimates established in points α) and β):

$$\begin{aligned}&\text{for } |x-x^0| \leqslant \varepsilon \min(r, r^m)(1 + |(x^0,t^0)|) \\ &\text{for } |t-t^0| = \varrho(1 + |(x^0,t^0)|) \\ &|P(x,t)| \geqslant |P(x^0,t)| - |P(x^0,t) - P(x,t)| \\ &\qquad \geqslant \frac{1}{2}\left(\frac{r}{4(m+1)}\right)^m (1 + |(x^0,t^0)|)^m .\end{aligned}$$

COROLLARY. *Assume that $P(x^0,t^0) = 0$ and let Σ denote the connected component of (x^0,t^0) in the set*

$$\{(x,t) \in \mathbb{C}^{k+1} | P(x,t) = 0\} \cap \{(x,t) \in \mathbb{C}^{k+1} | \, |x-x^0| < R(1 + |(x^0,t^0)|)\} .$$

Then on Σ we must have

$$|t - t^0| \leqslant \frac{1}{\varepsilon} \max(R, R^{1/m})(1 + |(x^0, t^0)|) .$$

PROOF. Set $\varepsilon \min(r, r^m) = R$ so that $r = \max\left(R/\varepsilon, (R/\varepsilon)^{1/m}\right)$ and $r \leqslant 1/\varepsilon \max(R, R^{1/m})$. By the previous lemma for some ϱ with $r/2 < \varrho < r$ we have that

$$\text{for } |x - x^0| \leqslant R(1 + |(x^0, t^0)|) \text{ and for } |t - t^0| = \varrho(1 + |(x^0, t^0)|)$$

we must have

$$|P(x, t)| > 0 .$$

Thus on Σ we must have

$$|t - t^0| < \varrho(1 + |(x^0, t^0)|) < \frac{1}{\varepsilon} \max(R, R^{1/m})(1 + |(x^0, t^0)|) .$$

Let V be an irreducible algebraic subvariety of $\mathbb{C}^n$ defined by a prime ideal $\mathfrak{b}$ in the ring $\mathcal{P}$ of polynomials in n variables, $\mathcal{P} = \mathbb{C}[z_1, \ldots, z_n]$. Let $d = \dim_{\mathbb{C}} V \geqslant 1$. Let $\mathfrak{a}$ be the asymptotic ideal of the ideal $\mathfrak{b}$ and let W be the corresponding asymptotic variety. Then W is a cone in $\mathbb{C}^n$ with vertex at the origin, purely dimensional of $\dim_{\mathbb{C}} W = d = \dim_{\mathbb{C}} V$.

We consider the decomposition of W into irreducible componentes

$$W = W_1 \cup \ldots \cup W_k .$$

For an algebraic subvariety $Z \subset \mathbb{C}^n$ we will denote by $\mathfrak{J}(Z)$ the ideal of $\mathcal{P}$ of polynomials vanishing on Z.

We know from algebraic geometry the following

LEMMA 8 (preparation lemma). *Let V and W be as above. After a real linear change of coordinates we may assume that the following conditions are satisfied*

(α) $\mathbb{C}[z_1, \ldots, z_d] \cap \mathfrak{J}(V) = 0$ *and* $\mathbb{C}[z_1, \ldots, z_d] \cap \mathfrak{J}(W_i) = 0$ *for* $1 \leqslant i \leqslant k$ *so that* $\mathbb{C}[z_1, \ldots, z_d]$ *can be considered as a subring of the rings*

$$\mathcal{P}/\mathfrak{J}(V) \quad \textit{and} \quad \mathcal{P}/\mathfrak{J}(W_i) \qquad \textit{for } 1 \leqslant i \leqslant k .$$

(β) *The rings* $\mathcal{P}/\mathfrak{J}(V)$ *and* $\mathcal{P}/\mathfrak{J}(W_i)$ *for* $1 \leqslant i \leqslant k$ *are integral over the ring* $\mathbb{C}[z_1, \ldots, z_d]$.

(γ) *For every j with $1 \leqslant j \leqslant n-d$ we can find a polynomial with principal part monic with respect to z_{d+j} in the ideals $\mathbb{C}[z_1, \ldots, z_d, z_{d+j}] \cap \mathfrak{J}(V)$ and $\mathbb{C}[z_1, \ldots, z_d, z_{d+j}] \cap \mathfrak{J}(W_i)$ for $1 \leqslant i \leqslant k$.*

(δ) *The ideals*

$$\mathbb{C}[z_1, \ldots, z_d, z_{d+1}] \cap \mathfrak{J}(V) = P(z_1, \ldots, z_{d+1})\mathbb{C}[z_1, \ldots, z_d, z_{d+1}]$$

and

$$\mathbb{C}[z_1, \ldots, z_d, z_{d+1}] \cap \mathfrak{J}(W_i) = Q_i(z_1, \ldots, z_{d+1})\mathbb{C}[z_1, \ldots, z_d, z_{d+1}]$$

for $1 \leqslant i \leqslant k$ are principal ideals with generators P and Q_i respectively, with principal part monic in z_{d+1}.

Moreover for $1 \leqslant i < j \leqslant k$ we have

$$\mathbb{C}[z_1, \ldots, z_d, z_{d+1}] \cap \mathfrak{J}(W_i) \neq \mathbb{C}[z_1, \ldots, z_d, z_{d+1}] \cap \mathfrak{J}(W_j) .$$

(ε) *The quotient field of the ring $\mathcal{P}/\mathfrak{J}(V)$ (resp. $\mathcal{P}/\mathfrak{J}(W_i)$ for $1 \leqslant i \leqslant k$) contains the field $\mathbb{C}(z_1, \ldots, z_d)$ of rational functions in the first d variables and is generated over it by the image of z_{d+1}.*

Moreover: if Δ is the discriminant of P with respect to z_{d+1}

if δ_i is the discriminant of Q_i with respect to z_{d+1}

we can find α_j and β_{ij} in $\mathbb{C}[z_1, \ldots, z_d, z_{d+1}]$ such that for any j with $2 \leqslant j \leqslant n-d$

$$\Delta z_{d+j} - \alpha_j \in \mathfrak{J}(V) .$$

$$\delta_i z_{d+j} - \beta_{ij} \in \mathfrak{J}(W_i) .$$

Therefore denoting by $\mathbb{C}^d$ the coordinate space of the first d coordinates and by $\omega: \mathbb{C}^n \to \mathbb{C}^d$ the natural projection we get locally biholomorphic surjective maps

$$\omega: V \cap \{\Delta \neq 0\} \to \mathbb{C}^d \cap \{\Delta \neq 0\}$$

$$\omega: W_i \cap \{\delta_i \neq 0\} \to \mathbb{C}^d \cap \{\delta_i \neq 0\} .$$

(ξ) *There exists a positive constant c such that*

$$|z| \leqslant c(1 + |(z_1, \ldots, z_d)|) \qquad \forall z \in V$$

and

$$|z| \leqslant c|(z_1, \ldots, z_d)| \qquad \forall z \in W_i \text{ for } 1 \leqslant i \leqslant k .$$

We can now give the proof of proposition 10

PROOF OF PROPOSITION 10. α) Let $\Theta \in W$; in view of lemma 6 it is enough to show that

$$g(\Theta) \leqslant \max \lim_{\substack{w \to \Theta \\ w \in W}} g(w) .$$

If $\Theta = 0$ this is true by the definition of the value $g(0)$. We may therefore assume $\Theta \neq 0$ and, as g is homogeneous of degree one, it is not restrictive to assume that $|\Theta| = 1$.

By the definition of $g(\Theta)$ we can find a sequence $\{z^{(\nu)}\} \subset V$ with

$$|z^{(\nu)}| > \nu , \qquad \frac{z^{(\nu)}}{|z^{(\nu)}|} \to \Theta$$

such that

$$g(\Theta) = \lim_{\nu \to \infty} \frac{\log |F(z^{(\nu)})|}{|z^{(\nu)}|} .$$

β) We may assume that V and W satisfy the conditions stated in the preparation lemma and we adopt the notations used there.

We denote by $D(z_1, \dots, z_d)$ the discriminant with respect to z_{d+1} of the polynomial $Q_1 Q_2 \dots Q_k$. We set

$$\Lambda(z_1, \dots, z_d) = \Delta(z_1, \dots, z_d) D(z_1, \dots, z_d) .$$

By a (real) linear change of coordinates $z_1, \dots, z_d$ in $\mathbb{C}^d$ we may as well assume that the principal part Λ^0 of the polynomial Λ is monic with respect to z_d. In fact such a linear change of coordinates does not affect the conditions of the preparation lemma.

Let μ be the degree of Λ (and thus of Λ^0) and set

$$x = (x_1, \dots, x_{d-1}) = (z_1, \dots, z_{d-1}) , \qquad t = z_d .$$

We now apply lemma 7 to the polynomial $\Lambda(x, t)$. We then can find a constant ε with $0 < \varepsilon < 1$ and for every $r > 0$ and for every choice of $(x^0, t^0) \in \mathbb{C}^d$ we can select ϱ with $r/2 < \varrho < r$ such that

$$\text{for any } x \text{ with } |x - x_0| \leqslant \varepsilon \min(r, r^m)(1 + |(x^0, t^0)|)$$

$$\text{for any } t \text{ with } |t - t^0| = \varrho(1 + |(x^0, t^0)|)$$

we have

$$(*) \qquad |\Lambda(x, t)| \geqslant \frac{1}{2}\left(\frac{r}{4(\mu + 1)}\right)^{\mu} (1 + |(x^0, t^0)|)^{\mu} .$$

γ) Let c be a constant such that

$$1 + |z| \leqslant c(1 + |(z_1, \dots, z_d)|)$$

for every point $z \in V$. This constant exists by virtue of the preparation lemma (point (ξ)).

We now remark that in every ideal $\mathbf{C}[z_1, \dots, z_d, z_{d+j}] \cap \mathfrak{J}(V)$ there exists a polynomial with principal part monic with respect to z_{d+j} for $1 \leqslant j \leqslant n-d$.

Let m be the maximal degree of these polynomials. To each one of them we can apply the corollary to lemma 7. Therefore there exists a constant σ with $0 < \sigma < 1$ such that, for any choice of $R > 0$ and of any point $z^0 = (z_1^0, \dots, z_n^0) \in V$, the following holds: let Σ denote the connected component of z^0 in the intersection

$$V \cap \left\{ z \in \mathbf{C}^n \,\middle|\, \left(\sum_1^d |z_i - z_i^0|^2\right)^{\frac{1}{2}} < \frac{R}{c}(1 + |z^0|) \right\}.$$

Then on Σ we have

$$(**) \qquad \sum_{d+1}^n |z_i - z_i^0| < \frac{1}{\sigma} \max(R, R^{1/m})(1 + |z^0|)\,.$$

δ) We now apply the considerations of points β) and γ) taking for z^0 any one $z^{(\nu)}$ of the points of the sequence $\{z^{(\nu)}\}$ defined in point α).

We fix r with $0 < r < 1$, for every ν we can find ϱ_ν with $r/2 < \varrho_\nu < r$ so that the condition $(*)$ is satisfied for the point $(x^0, t^0) = (z_1^{(\nu)}, \dots, z_{d-1}^{(\nu)}, z_d^{(\nu)})$.

We consider in $\mathbf{C}^d$ the sequence of one-dimensional discs

$$D_\nu(r) =$$
$$= \left\{(z_1, \dots, z_d) \in \mathbf{C}^d \,|\, z_1 = z_1^{(\nu)}, \dots, z_{d-1} = z_{d-1}^{(\nu)}, |z_d - z_d^{(\nu)}| < r(1 + |(z_1^{(\nu)}, \dots, z_d^{(\nu)})|)\right\}.$$

If $\pi\colon V \to \mathbf{C}^d$ denotes the natural projection of V onto $\mathbf{C}^d$ (i.e. with the notation of the preparation lemma $\pi = \omega|V$) then

$$z^{(\nu)} \in \pi^{-1}(D_\nu(r))$$

and we can consider the connected component $\Sigma^{(\nu)}(r)$ of $z^{(\nu)}$ in $\pi^{-1}(D_\nu(r))$. Then $\Sigma^{(\nu)}(r)$ is a Riemann surface (possibly reducible) over the disc $D^{(\nu)}(r)$.

Taking in point γ) $R/c = r$ because of $(**)$ and because $r < 1$ we do have for any point $v^{(\nu)} \in \Sigma^{(\nu)}(r)$

$$\text{i)} \qquad \sum_{d+1}^n |v_i^{(\nu)} - z_i^{(\nu)}| < \frac{r^{1/m}}{\sigma}(1 + |z^{(\nu)}|)\,.$$

In particular the point $v^{(\nu)}$ can be so chosen that

ii) $|v_d^{(\nu)} - z_d^{(\nu)}| = \varrho_\nu(1 + |(z_1^{(\nu)}, \dots, z_d^{(\nu)})|)$

iii) $v_j^{(\nu)} = z_j^{(\nu)}$ for $1 \leqslant j \leqslant d-1$.

By the maximum principle applied to the holomorphic function $F|\Sigma^{(\nu)}(r)$ we can also satisfy the condition

iv) $|F(v^{(\nu)})| \geqslant |F(z^{(\nu)})|$.

Finally because of condition $(*)$ in point α) we must have

$$\text{v)} \qquad |\Delta(v_1^{(\nu)}, \dots, v_d^{(\nu)})| \geqslant \frac{1}{2}\left(\frac{r}{4(\mu+1)}\right)^\mu (1 + |(z_1^{(\nu)}, \dots, z_d^{(\nu)})|)^\mu .$$

ε) Because of properties i), ii) and iii) the subsequence of $\mathbb{C}^n$ given by $\{v^{(\nu)}/|z^{(\nu)}|\}$ is a bounded sequence. Passing to a subsequence, that we still denote by $\{v^{(\nu)}/|z^{(\nu)}|\}$, we may assume that the sequence is convergent:

$$\frac{v^{(\nu)}}{|z^{(\nu)}|} \to w .$$

Since $|z^{(\nu)}| \to \infty$ the limit point $w \in W$.

From condition i) dividing by $|z^{(\nu)}|$ and passing to the limit for $\nu \to \infty$ we obtain

$$(a) \qquad \sum_{d+1}^{n} |w_i - \Theta_i| \leqslant \frac{r^{1/m}}{\sigma} .$$

From condition ii) we get

$$\frac{r}{2}\,\frac{1 + |(z_1^{(\nu)}, \dots, z_d^{(\nu)})|}{|z^{(\nu)}|} < \left|\frac{v_d^{(\nu)}}{|z^{(\nu)}|} - \frac{z_d^{(\nu)}}{|z^{(\nu)}|}\right| < r\,\frac{1 + |(z_1^{(\nu)}, \dots, z_d^{(\nu)})|}{|z^{(\nu)}|}$$

and therefore

$$(b) \qquad \frac{r}{2}|(\Theta_1, \dots, \Theta_d)| \leqslant |w_d - \Theta_d| \leqslant r|(\Theta_1, \dots, \Theta_d)| .$$

From condition iii) we get

$$(c) \qquad w_j = \Theta_j \qquad \text{for } 1 \leqslant j \leqslant d-1 .$$

Since $|\Theta| = 1$, because of the last condition of the preparation lemma, we must have $|(\Theta_1, \dots, \Theta_d)| > 0$. Therefore from condition v) we derive the inequality

$$|\Lambda^0(w_1, \ldots, w_d)| \geqslant \frac{1}{2}\left(\frac{r}{4(\mu+1)}\right)^{\mu} |(\Theta_1, \ldots, \Theta_d)|^{\mu} > 0 .$$

Since $D(z_1, \ldots, z_d)$ is a homogeneous polynomial, it must be a factor of the principal part $\Lambda_0(z_1, \ldots, z_d)$ of Λ. The above inequality shows therefore that w is a non singular point of W.

Finally from condition iv) we derive that

$$g(w) \geqslant \max_{\nu\to\infty} \lim \frac{\log |F(v^{(\nu)})|}{|z^{\nu}|} \geqslant \max_{\nu\to\infty} \lim \frac{\log |F(z^{(\nu)})|}{|z^{(\nu)}|} = g(\Theta) .$$

As r with the condition $0 < r < 1$ can be chosen arbitrarily small we deduce that w can be chosen in an arbitrarily small neighborhood of Θ as it follows from the relations (a) (b) and (c).

We have thus proved the following

STATEMENT. *Given $\Theta \in W$ with $|\Theta| = 1$ and any $\beta > 0$ we can find a non singular point $w \in W$ with*

$$|w - \Theta| < \beta ,$$

$$g(w) \geqslant g(\Theta) .$$

The above statement implies that

$$\max_{\substack{w\to\Theta \\ w\in W}} \lim g(w) \geqslant g(\Theta) .$$

This is what we wanted to prove.

REMARK. Let $S(W)$ denote the subset of W of singular points in W. We have also proved with the previous argument that given any $\Theta \in W$ we have

$$g(\Theta) = \max_{\substack{w\to\Theta \\ w\in W-S(W)}} \lim g(w) .$$

More precisely if we denote by A the algebraic subset of W defined by $A = \{w \in W | \Lambda^0(w) = 0\}$ then A has codimension $\geqslant 1$ on any irreducible component of W and moreover $A \supset S(W)$. What we have proved is that for any $\Theta \in W$

$$g(\Theta) = \max_{\substack{w\to\Theta \\ w\in W-A}} \lim g(w) .$$

PROPOSITION 11. *Let F be a holomorphic function of exponential type defined on the irreducible algebraic variety $V \subset \mathbb{C}^n$. Let g denote the gauge function associated to F and defined on the asymptotic variety W of V. Let $S(W)$ denote the subset of singular points of W.*

Then g is a plurisubharmonic function at any point $\Theta \in W - S(W)$.

Before giving the proof of this proposition we establish two lemmas. We may assume that V and W satisfy the conditions stated in the preparation lemma 8. We adopt the notations introduced in that lemma.

Let D denote the discriminant with respect to z_{d+1} of the polynomial $Q_1 \dots Q_k$ and set $\Lambda(z_1, \dots, z_d) = \Delta D$. We denote by Λ^0 the principal part of Λ.

Let $\mathbb{C}^d$ (resp. $\mathbb{C}^{n-d}$) be the coordinate subspace of $\mathbb{C}^n$ of the first d coordinates (resp. of the last $n-d$ coordinates) so that $\mathbb{C}^n = \mathbb{C}^d \times \mathbb{C}^{n-d}$.

Let $\Theta = (\alpha, \beta) \in \mathbb{C}^d \times \mathbb{C}^{n-d}$ with

$$\alpha = (\Theta_1, \dots, \Theta_d) \in \mathbb{C}^d \quad \text{and } \beta = (\Theta_{d+1}, \dots, \Theta_n) \in \mathbb{C}^{n-d} .$$

For $r_1 > 0$, $r_2 > 0$ we set

$$B(\alpha, r_1) = \{\eta \in \mathbb{C}^d \,|\, |\eta - \alpha| < r_1\}$$

$$B(\Theta, r_1, r_2) = \{(\eta, w) \in \mathbb{C}^d \times \mathbb{C}^{n-d} \,|\, |\eta - \alpha| < r_1, \ |w - \beta| < r_2\}$$

where as usual by $|\cdot|$ we denote the euclidean norm.

LEMMA 9. *Let $\Theta = (\alpha, \beta) \in W$ and let $\Lambda^0(\alpha) \neq 0$.*
We can choose positive numbers $r_1 > 0$, $r_2 > 0$, $\lambda_0 > 0$ such that

i) *The set $W \cap B(\Theta, r_1, r_2)$ is the graph of a holomorphic function*

$$h\colon B(\alpha, r_1) \to \mathbb{C}^{n-d}$$

with

$$h(\alpha) = \beta \qquad |h(\eta) - \beta| < r_2 \qquad \textit{for } \eta \in B(\alpha, r_1);$$

$$W \cap B(\Theta, r_1, r_2) = \{(\eta, h(\eta)) \in \mathbb{C}^d \times \mathbb{C}^{n-d} | \eta \in B(\alpha, r_1)\} .$$

ii) *Set*

$$\Gamma = \bigcup_{\lambda \geq \lambda_0} \lambda B(\Theta, r_1, r_2)$$

$$\gamma = \bigcup_{\lambda \geq \lambda_0} \lambda B(\alpha, r_1) .$$

There exist an integer $s \geqslant 1$ *and* s *holomorphic functions*

$$f_j: \gamma \to \mathbf{C}^{n-d} \qquad \text{for } 1 \leqslant j \leqslant s$$

such that

(α) *for every* $\eta \in \gamma$, $(\eta, f_j(\eta)) \in \Gamma$ *for* $1 \leqslant j \leqslant s$.

(β) *the set* $V \cap \Gamma$ *decomposes into the union of the graphs of these* s *functions:*

$$V \cap \Gamma = \bigcup_{j=1}^{s} \{(\eta, f_j(\eta)) \in \mathbf{C}^d \times \mathbf{C}^{n-d} | \eta \in \gamma\}$$

iii) *For every* $\eta \in \gamma$ *we have* $f_i(\eta) \neq f_j(\eta)$ *if* $i \neq j$ $1 \leqslant i, j \leqslant s$.

iv) *For any* j *with* $1 \leqslant j \leqslant s$ *and for* $\beta > 0$ *we have*

$$\lim_{\lambda \to +\infty} \frac{1}{\lambda} f_j(\lambda \eta) = h(\eta) \qquad \forall \eta \in B(\alpha, r_1)$$

and the convergence is uniform on compact subsets of $B(\alpha, r_1)$.

Proof of i). Since $Q_1, \ldots, Q_k$ are homogeneous polynomials, the discriminant D of $Q_1 \ldots Q_k$ is also homogeneous and therefore it is a factor of Δ^0. Hence $D(\alpha) \neq 0$. Therefore also (with the notations of the preparation lemma) $\delta_i(\alpha) \neq 0$ for $1 \leqslant i \leqslant k$ as each δ_i is a factor of D.

Let

$$\Theta_{d+1} = \Theta_{d+1}^{(1)}, \Theta_{d+1}^{(2)}, \ldots, \Theta_{d+1}^{(l)}$$

denote the simple distinct roots of

$$(Q_1 \ldots Q_k)(\alpha, t) = 0 .$$

We choose $\varepsilon > 0$ so small that on the t-plane the l discs

$$|t - \Theta_{d+1}^{(j)}| \leqslant 2\varepsilon , \quad 1 \leqslant j \leqslant l ,$$

are disjoint. We then can find $\sigma > 0$ such that for

$$|\eta - \alpha| < \sigma$$

the equation $(Q_1 \ldots Q_k)(\eta, t) = 0$ admits a unique root $t = \tau_j(\eta)$ in the disc $|t - \Theta_{d+1}^{(j)}| < \varepsilon$ for $1 \leqslant j \leqslant l$. Since $Q_1 \ldots Q_k$ is monic in t these are the only roots of $(Q_1 \dots Q_k)(\eta, t) = 0$. One verifies by means of the logarithmic integral that the l functions $\tau_j(\eta)$ are holomorphic for $|\eta - \alpha| < \sigma$. We may also assume that for $|\eta - \alpha| < \sigma$ we have $\Lambda^0(\eta) \neq 0$ and that $\tau_1(\alpha) = \Theta_{d+1}$.

For $1 \leqslant j \leqslant l$ and for $|\eta - \alpha| < \sigma$ the root $\tau_j(\eta)$ is the root of only one, $Q_{s(j)}$ say, of the polynomials $Q_1, \dots, Q_k$. To each one $\tau_j(\eta)$ of these roots corresponds therefore a unique point $w^{(j)}(\eta) \in W$

$$w^{(j)}(\eta) = \begin{cases} w_1 = \eta_1, \dots, w_d = \eta_d, \qquad w_{d+1} = \tau_j(\eta) \\ w_{d+2} = \delta_s^{-1}(\eta)\beta_{s2}(\eta, \tau_j(\eta)), \dots, w_n = \delta_s^{-1}(\eta)\beta_{s,n-d}(\eta, \tau_j(\eta)) . \end{cases}$$

These l points $w^{(j)}(\eta)$ are distinct for every η with $|\eta - \alpha| < \sigma$.

By construction the l balls

$$\left\{\sum_{h=1}^{n-d} |w_{d+h} - w_{d+h}^{(j)}(\alpha)|^2\right\}^{\frac{1}{2}} < \varepsilon \qquad 1 \leqslant j \leqslant l$$

are disjoint. If σ is sufficiently small for any η with $|\eta - \alpha| < \sigma$ the point $(w_{d+1}^{(j)}(\eta), \dots, w_n^{(j)}(\eta))$ is contained in the j-th of the above balls.
We can thus take $r_1 = \sigma$, $r_2 = \varepsilon$ and define h by

$$h(\eta) = \left(\tau_1(\eta), \delta_{s(1)}^{-1}(\eta)\beta_{s(1),2}(\eta, \tau_1(\eta)), \dots, \delta_{s(1)}^{-1}(\eta)\beta_{s(1),n-d}(\eta, \tau_1(\eta))\right) .$$

We note explicitly that we can choose $r_2 = \varepsilon$ arbitrarily small provided we choose $r_1 = \sigma$ sufficiently small.

Proof of ii), iii) and iv). We divide the proof in several steps

α) Let P^0 denote the principal part of the polynomial P in the preparation lemma. We want to show that

we can choose $r_1 > 0$, $r_2 > 0$ and $\delta > 0$ such that for $\eta \in B(\alpha, r_1)$ and t such that $|t - \Theta_{d+1}| = r_2$ we have

$$|P^0(\eta, t)| > \delta^m$$

where m is the degree of P^0.

In fact for some integers $s_1 \geqslant 1, \dots, s_k \geqslant 1$ we will have

$$P^0 = Q_1^{s_1} Q_2^{s_2} \dots Q_k^{s_k} .$$

Therefore, for $|\eta - \alpha| < \sigma$, $t = \tau_j(\eta)$ for $1 \leqslant j \leqslant l$ are the only distinct roots of the equation $P^0(\eta, t) = 0$.

Now we have

$$|\tau_j(\eta) - \tau_1(\eta)| > 2\varepsilon \quad \text{if } j > 1$$

and

$$|\tau_1(\eta) - \Theta_{d+1}| < \varepsilon/2 \quad \text{if } |\eta - \alpha| < \sigma$$

provided σ is chosen sufficiently small. Therefore for

$$|\eta - \alpha| < \sigma \quad \text{and } |t - \Theta_{d+1}| = \varepsilon$$

we obtain:

for $j > 1$

$$\begin{aligned}|t - \tau_j(\eta)| &\geqslant |\tau_1(\eta) - \tau_j(\eta)| - |t - \Theta_{d+1}| - |\Theta_{d+1} - \tau_1(\eta)| \\ &> 2\varepsilon - \varepsilon - \varepsilon/2 \\ &\geqslant \varepsilon/2\end{aligned}$$

and for $j = 1$

$$\begin{aligned}|t - \tau_1(\eta)| &\geqslant |t - \Theta_{d+1}| - |\tau_1(\eta) - \Theta_{d+1}| \\ &> \varepsilon - \varepsilon/2 \\ &\geqslant \varepsilon/2 .\end{aligned}$$

As P^0 is monic in t we have $P^0(\eta, t) = \prod_{j=1}^{m} (t - \tau_j(\eta))$ each root being repeated with the proper multiplicity. If we take $r_1 = \sigma$, $r_2 = \varepsilon$ and $\delta = \varepsilon/2$ we obtain the desired inequality.

β) We have $\Lambda_0(\eta) \neq 0$ for $\eta \in B(\alpha, r_1)$. We want now to show that:

We can find $\lambda_0 > 0$ *such that for*

$$\eta \in \gamma = \bigcup_{\lambda \geqslant \lambda_0} \lambda B(\alpha, r_1)$$

we have $\Delta(\eta) \neq 0$.

Let Δ^0 denote the principal part of Δ and let μ be the degree of Δ. As Δ^0 is a factor of Λ^0 we have $\Delta^0(\eta) \neq 0$ for $\eta \in B(\alpha, r_1)$.

Replacing r_1 with a smaller positive quantity we may assume that

$$\inf_{\eta \in B(\alpha, r_1)} |\Delta_0(\eta)| > \varrho > 0$$

for some $\varrho > 0$. Therefore for $\eta \in \bigcup_{\lambda>0} \lambda B(\alpha, r_1)$ we obtain an inequality of the form

$$|\Delta^0(\eta)| > c|\eta|^\mu$$

for some constant $c > 0$.

From the other hand, since $\Delta - \Delta^0$ is a polynomial of degree $\leqslant \mu - 1$ we obtain, with some constant $c_1 > 0$, an estimate, for every $\eta \in \mathbf{C}^d$,

$$|\Delta(\eta) - \Delta^0(\eta)| \leqslant c_1(1 + |\eta|)^{\mu-1}.$$

Therefore for $\eta \in \gamma$ we have

$$\begin{aligned} |\Delta(\eta)| &\geqslant |\Delta_0(\eta)| - |\Delta(\eta) - \Delta^0(\eta)| \\ &\geqslant c|\eta|^\mu - c_1(1 + |\eta|)^{\mu-1} \\ &\geqslant \frac{c}{2}|\eta|^\mu > 0 \end{aligned}$$

provided $|\eta| > c_1/c_2$ i.e. provided $\lambda_0 > c_1/2c(|\alpha| - r_1)$. Note that since $\Delta^0(\eta) \neq 0$ on $B(\alpha, r_1)$ we have $0 \notin B(\alpha, r_1)$, thus $|\alpha| > r_1$.

γ) The root $t = \tau_1(\eta)$ of the equation $P^0(\eta, t) = 0$ for $\eta \in B(\alpha, r_1)$ will have a constant multiplicity $s = s_j$ for some j with $1 \leqslant j \leqslant k$ since $D(\eta) \neq 0$ on $B(\alpha, r_1)$.

For $\lambda > 0$ we consider the polynomial of degree $\leqslant m - 1$ in η and t

$$P^0(\eta, t) - \lambda^{-m} P(\lambda\eta, \lambda t).$$

The coefficients of this polynomial will decrease as $1/\lambda$ when $\lambda \to \infty$. Therefore for

$$\eta \in B(\alpha, r_1), \quad |t - \Theta_{d+1}| \leqslant r_2$$

we can find a constant $c > 0$ such that, for $\lambda > 1$ we have

$$|P^0(\eta, t) - \lambda^{-m} P(\lambda\eta, \lambda t)| < \frac{c}{\lambda}.$$

If we choose $\lambda > 2c/\delta^m$ we will have for

$$\eta \in B(\alpha, r_1), \quad |t - \Theta_{d+1}| = r_2,$$

that

$$|P^0(\eta, t) - \lambda^{-m} P(\lambda\eta, \lambda t)| < \delta^m < |P^0(\eta, t)|.$$

For any $\lambda > \max(1, 2c/\delta^m, \lambda_0) = \lambda_1$, we will have

for any $\eta \in B(\alpha, r_2)$ that $\Delta(\lambda\eta) \neq 0$

and, by the theorem of Rouché, that

for any $\eta \in B(\alpha, r_1)$ the equation $P(\lambda\eta, \lambda t) = 0$ has exactly s distinct roots in t in the disc $|t - \Theta_{d+1}| < r_2$;

$$\varphi_1(\lambda, \eta), \ldots, \varphi_s(\lambda, \eta) .$$

Note that, for $1 \leqslant i \leqslant s$,

$$\lambda\varphi_i(\lambda, \eta) = \psi_i(\lambda\eta)$$

is a holomorphic function of $\lambda\eta$ for $\eta \in B(\alpha, r_1)$ and any $\lambda > \lambda_1$. For each i with $1 \leqslant i \leqslant s$ we get then a unique point $v(\lambda\eta) \in V$ for

$$\eta \in B(\alpha, r_1) \qquad \text{and } \lambda > \lambda_1$$

$$v(\lambda\eta) \equiv \begin{cases} z_1 = \lambda\eta_1, \ldots, z_d = \lambda\eta_d, & z_{d+1} = \psi_i(\lambda\eta) \\ z_{d+2} = \Delta(\lambda\eta)^{-1}\alpha_2(\lambda\eta, \psi_i(\lambda\eta)), \ldots, z_n = \Delta(\lambda\eta)^{-1}\alpha_{n-d}(\lambda\eta, \psi_i(\lambda\eta)) . \end{cases}$$

We set, for $\eta \in \bigcup_{\lambda \geqslant \lambda_1} \lambda B(\alpha, r_1)$ and $1 \leqslant i \leqslant s$,

$$f_i(\lambda\eta) \equiv \{\psi_i(\lambda\eta), \Delta(\lambda\eta)^{-1}\alpha_2(\lambda\eta, \psi_i(\lambda\eta)), \ldots, \Delta(\lambda\eta)^{-1}\alpha_{n-d}(\lambda\eta, \psi_i(\lambda\eta))\}$$

so that setting $\gamma_1 = \bigcup_{\lambda > \lambda_1} \lambda B(\alpha, r_1)$ we have

$$f_i\colon \gamma_1 \to \mathbf{C}^{n-d}$$

holomorphic and for $i \neq j$ and $\eta \in \gamma_1$ $f_i(\eta) \neq f_j(\eta)$.

δ) We fix i with $1 \leqslant i \leqslant s$ and we consider for $\lambda > \lambda_1$ the holomorphic functions (depending on the parameter λ),

$$\frac{\psi_i(\lambda\eta)}{\lambda} = \frac{\psi(\lambda\eta)}{\lambda} \qquad \text{for } \eta \in B(\alpha, r_1) .$$

We have by construction

$$\left|\frac{\psi(\lambda\eta)}{\lambda}\right| < r_2 + |\Theta_{d+1}|$$

for any $\lambda > \lambda_1$ so that those functions are equibounded for $\eta \in B(\alpha, r_2)$.

Also set

$$\lambda^{-m}P(\lambda\eta, \lambda t) = P^0(\eta, t) + \lambda^{-m}Q(\lambda\eta, \lambda t)$$

and choose any sequence $\{\lambda_h\}_{h\in\mathbb{N}}$ with $\lambda_h > \lambda_1$ and $\lambda_h \to +\infty$.

We have

$$P^0\left(\eta, \frac{\psi(\lambda_h\eta)}{\lambda_h}\right) + \lambda_h^{-m}Q(\lambda_h\eta, \lambda_h\psi(\lambda_h\eta)) = 0 .$$

Therefore for $h \to +\infty$ we must have for $\eta \in B(\alpha, r_1)$

$$\lim_{h\to\infty} \frac{\psi(\lambda_h\eta)}{\lambda_h} = \tau_1(\eta)$$

because for $\eta \in B(\alpha, r_1)$, $\lim_{h\to\infty} \lambda_h^{-m}Q(\lambda_h\eta, \lambda_h\psi(\lambda_h\eta)) = 0$.

We set (cf. proof of i)) for $\eta \in B(\alpha, r_1)$

$$h(\eta) = (h_{d+1}(\eta), \ldots, h_n(\eta)) ,$$

we have proved that for any sequence $\{\lambda_h\}_{h\in\mathbb{N}}$ with $\lambda_h > \lambda_1$ and $\lambda_h \to +\infty$ we must have

$$\lim_{h\to\infty} \frac{\psi(\lambda_h\eta)}{\lambda_h} = h_{d+1}(\eta)$$

for any $\eta \in B(\alpha, r_2)$ and uniformly on compact subsets of $B(\alpha, r_1)$.

ε) We set

$$f_i(\lambda\eta) = f(\lambda\eta) = (f_{d+1}(\lambda\eta), \ldots, f_n(\lambda\eta)) .$$

From condition ξ) in the preparation lemma we obtain

$$\left|\frac{f_{d+j}(\lambda\eta)}{\lambda}\right| \leqslant c\,\frac{(1 + |\lambda\eta|)}{\lambda}, \qquad 1 \leqslant j \leqslant n-d ,$$

for $\lambda > \lambda_1$ and $\eta \in B(\alpha, r_1)$ and therefore for $\lambda > \lambda_1$ and $\eta \in B(\alpha, r_1)$ the holomorphic functions $f_{d+j}(\lambda\eta)/\lambda$ are equibounded. For every $\eta \in B(\alpha, r_1)$ fixed and for a sequence $\{\lambda_h\}_{h\in\mathbb{N}}$ with $\lambda_h > \lambda_1$ and $\lambda_h \to +\infty$ such that there exists the limit for any j, $1 \leqslant j \leqslant n-d$

$$\lim_{h\to\infty} \frac{f_{d+j}(\lambda_h\eta)}{\lambda_h}$$

we have:
for any polynomial $p(z)$ vanishing on V we must have

$$\lambda_h^{-\deg p} p(\lambda_h \eta, f(\lambda_h \eta)) = 0$$

and therefore at the limit, for $h \to \infty$, we must have

$$p^0\left(\eta, \lim_{h\to\infty} \frac{f(\lambda_h \eta)}{\lambda_h}\right) = 0 .$$

This shows that the point $w = (\eta, \lim_{h\to\infty} f(\lambda_h\eta)/\lambda_h)$ is in W, $w \in W$. But $f_{d+1}(\lambda\eta) = \psi(\lambda\eta)$ and we know from point δ) that $\lim_{h\to\infty} \psi(\lambda_h\eta)/\lambda_h = h_{d+1}(\eta)$. This shows that the point w must be the point $(\eta, h(\eta)) \in W$.

This implies that for any $\eta \in B(\alpha, r_1)$ there exists the limit

$$\lim_{\lambda\to\infty} \frac{f(\lambda\eta)}{\lambda} = h(\eta) .$$

Since the functions $f_{d+j}(\lambda\eta)/\lambda$ are equibounded for $\eta \in B(\alpha, r_1)$ that convergence must be uniform on any compact subset $K \subset B(\alpha, r_1)$. Points iii) and iv) of the lemma have been therefore established.

ξ) It remains to complete the proof of point ii). For this purpose we choose r_1' with $0 < r_1' < r_1$ so small that for $\eta \in B(\alpha, r_1')$ we have $|h(\eta) - \beta| < r_2/2$. Then we choose $\lambda_3 > 0$ such that for $\lambda \geqslant \lambda_3$ we have

$$\left|\frac{1}{\lambda} f_i(\lambda\eta) - h(\eta)\right| < r_2/2$$

for $\eta \in B(\alpha, r_1')$ and $1 \leqslant i \leqslant s$.

Then for

$$\eta \in \gamma_2 = \bigcup_{\lambda \geqslant \lambda_3} \lambda B(\alpha, r_1')$$

we have for $1 \leqslant i \leqslant s$

$$(\eta, f_i(\eta)) \in \bigcup_{\lambda\geqslant\lambda_3} \lambda B(\Theta, r_1', r_2) .$$

Therefore setting

$$\Gamma = \bigcup_{\lambda\geqslant\lambda_3} \lambda B(\Theta, r_1', r_2)$$

we have

$$V \cap \Gamma = \bigcup_{j=1}^{s} \{(\eta, f_j(\eta)) \in \mathbb{C}^d \times \mathbb{C}^{n-d} | \eta \in \gamma_2\} .$$

This completes the proof as we have only to replace in the notations $\lambda_3, r_1', \gamma_2$ with λ_0, r_1, γ.

LEMMA 10. *Let Ω be an open set in $\mathbb{C}^n$ and let $\{\varphi_\lambda\}_{\lambda \geq \lambda_0}$ be a family of plurisubharmonic functions defined on Ω.*

We assume that on any compact subset of Ω the family $\{\varphi_\lambda\}_{\lambda \geq \lambda_0}$ is uniformly bounded from above.

Define for $z \in \Omega$

$$\varphi(z) = \max \lim_{\substack{w \to z \\ \lambda \to \infty}} \varphi_\lambda(w) .$$

Then φ is plurisubharmonic on Ω.

PROOF. First we prove that φ is upper-semicontinuous. Given $z \in \Omega$ and $\varepsilon > 0$ we can find $\lambda_1 \geq \lambda_0$ and $\delta > 0$ such that

for $\lambda \geq \lambda_1$ and $w \in \Omega$ with $|w - z| < \delta$

we have

$$\varphi_\lambda(w) \leq \varphi(z) + \varepsilon .$$

Therefore for any $w_1 \in \Omega$ with $|w_1 - z| < \delta/2$ we get

$$\varphi(w_1) \leq \varphi(z) + \varepsilon .$$

This shows that φ is upper semicontinuous.

To prove that φ is plurisubharmonic we show that for any $z_0 \in \Omega$, for any $a \in \mathbb{C}^n$ with $|a| = 1$, for any $\varepsilon > 0$ with $\varepsilon < \operatorname{dist}(z_0, \partial\Omega)$ we have

$$\varphi(z_0) \leq \frac{1}{2\pi} \int_0^{2\pi} \varphi(z_0 + a\varepsilon \exp(it))\, dt .$$

Now we have

$$\begin{aligned}
\varphi(z_0) &= \max \lim_{\substack{w \to z_0 \\ \lambda \to \infty}} \varphi_\lambda(w) \\
&\leq \max \lim_{\substack{w \to z_0 \\ \lambda \to \infty}} \frac{1}{2\pi} \int_0^{2\pi} \varphi_\lambda(w + a\varepsilon \exp(it)) dt \\
&\leq \frac{1}{2\pi} \int_0^{2\pi} \max \lim_{\substack{w \to z_0 \\ \lambda \to \infty}} \varphi_\lambda(w + a\varepsilon \exp(it))\, dt \qquad \text{(by Fatou's lemma)} \\
&= \frac{1}{2\pi} \int_0^{2\pi} \varphi(z_0 + a\varepsilon \exp(it))\, dt .
\end{aligned}$$

PROOF OF PROPOSITION 11. We first show that g is plurisubharmonic at any point $\Theta = (\alpha, \beta) \in W$ with $\Lambda^0(\alpha) \neq 0$. We use here the notations of lemma 9.

Set for $\eta \in B(\alpha, r_1)$

$$\varphi_\lambda(\eta) = \sup_{1 \leqslant i \leqslant s} \frac{\log |F(\lambda\eta, f_i(\lambda\eta))|}{\lambda} \qquad \text{for } \lambda \geqslant \lambda_0 .$$

Then $\varphi_\lambda(\eta)$ is plurisubharmonic in $B(\alpha, r_1)$. Moreover the functions φ_λ are uniformly bounded from above because F is on V of exponential type. We have indeed, taking into account point ξ) of the preparation lemma, for some constant $c > 0$,

$$\log |F(\lambda\eta, f_i(\lambda\eta))| \leqslant \log c + c|\lambda\eta| .$$

We can therefore apply lemma 10. We obtain that

$$\text{for } \eta \in B(\alpha, r_1),\ \varphi(\eta) = \max_{\substack{y \to \eta \\ \lambda \to \infty}} \lim \varphi_\lambda(y)$$

is a plurisubharmonic function.

We will show that

$$\varphi(\eta) = g(\eta, h(\eta)) \qquad \forall \eta \in B(\alpha, r_1)$$

and this will prove the subharmonicity of g where $\Lambda^0 \neq 0$.

Choose sequences $\{\lambda_h\}_{h \in \mathbf{N}}$ and $\{y_h\}_{h \in \mathbf{N}}$ with $\lambda_h \to +\infty$ $\lambda_h > \lambda_0$ and with $y_h \in B(\alpha, r_1)$ $y_h \to \eta$ with $\eta \in B(\alpha, r_1)$, and assume that

$$\varphi(\eta) = \lim_{h \to \infty} \varphi_{\lambda_h}(y_h) .$$

Consider the sequence of points

$$z_h = (\lambda_h y_h, f_i(\lambda_h y_h))$$

for a fixed i with $1 \leqslant i \leqslant s$. We have,

$$z_h \in V , \qquad |z_h| \to \infty , \qquad \frac{z_h}{|z_h|} \to \frac{(\eta, h(\eta))}{|(\eta, h(\eta))|} .$$

We may assume that i is so chosen that

$$\varphi(\eta) = \lim_{h \to \infty} \frac{\log |F(\lambda_h y_h, f_i(\lambda_h y_h))|}{\lambda_h} .$$

Thus

$$\begin{aligned}\varphi(\eta) &= \lim_{h\to\infty} \frac{\log|F(z_h)|}{|z_h|}\cdot\frac{|z_h|}{\lambda_h}\\ &= \lim_{h\to\infty} \frac{\log|F(z_h)|}{|z_h|}\cdot|(\eta, h(\eta))|\\ &\leq g(\eta, h(\eta)) .\end{aligned}$$

We have to show the opposite inequality

$$g(\eta, h(\eta)) \leq \varphi(\eta) .$$

We choose a sequence $\{(a_h, b_h)\}_{h\in\mathbf{N}} \subset V$ with

$$|(a_h, b_h)| \to \infty , \qquad \frac{(a_h, b_h)}{|(a_h, b_h)|} \to \frac{(\eta, h(\eta))}{|(\eta, h(\eta))|}$$

for some $\eta \in B(\alpha, r_1)$, so that

$$g(\eta, h(\eta)) = \lim_{h\to\infty} \frac{\log|F(a_h, b_h)|}{|a_h, b_h)|}\cdot|(\eta, h(\eta))| .$$

We set

$$\lambda_h = \frac{|(a_h, b_h)|}{|(\eta, h(\eta))|}$$

so that $\lambda_h \to +\infty$. We then define, for h sufficiently large,

$$y_h = \frac{a_h}{\lambda_h} = \frac{|(\eta, h(\eta))|}{|a_h, b_h)|}\cdot a_h .$$

Thus for $h \to \infty$ we have $y_h \to \eta \in B(\alpha, r_1)$. We also have that, for large h,

$$\frac{b_h}{\lambda_h} = \frac{b_h}{|(a_h, b_h)|}\cdot|(\eta, h(\eta))| \to h(\eta) .$$

Therefore for large h we must have

$$b_h = f_{i_h}(\lambda_h y_h)$$

for some i_h with $1 \leq i_h \leq s$.

We can then write

$$
\begin{aligned}
g(\eta, h(\eta)) &= \lim_{h\to\infty} \frac{\log|F(a_h, b_h)|}{|(a_h, b_h)|/|(\eta, h(\eta))|} \\
&= \lim_{h\to\infty} \frac{\log|F(\lambda_h y_h, f_{i_h}(\lambda_h y_h))|}{\lambda_h} \\
&\leqslant \max \lim_{h\to\infty} \varphi_{\lambda_h}(y_h) \\
&\leqslant \varphi(\eta)\,.
\end{aligned}
$$

We conclude therefore that on $W \cap \{\Lambda^0 \neq 0\}$ the gauge function g is plurisubharmonic.

Now the gauge function g is locally bounded on W from above because F is a holomorphic function of exponential type. Let $\Theta \in W \cap \{\Lambda^0 = 0\}$ be a non singular point of W. By the remark at the end of the proof of proposition 10 we have

$$
g(\Theta) = \max \lim_{\substack{w\to\Theta \\ w\in W \cap \{\Lambda_0 \neq 0\}}} g(w)\,.
$$

This shows that the value of g at Θ is the value of the plurisubharmonic extension of $g|W \cap \{\Lambda^0 \neq 0\}$ given by the analog of the Riemann extension theorem. Hence g is plurisbharmonic also at the point Θ (cf. [10] Sastz 6 p. 183).

One can also argue as follows. Given $\Theta \in W$ non singular, the coordinates in the preparation lemma 8 can always be so chosen that $\Theta \notin W \cap \{\Lambda^0 = 0\}$. Thus for any non singular point $\Theta \in W$ we can argue as we have done above.

Let us denote as before by W the asymptotic variety of the irreducible characteristic variety V and let

$$
W = W_1 \cup \ldots \cup W_k
$$

be a decomposition of W into irreducible components.

Let $W_i^* \xrightarrow{\nu_i} W_i$ denote the normalization of the variety W_i so that the disjoint union

$$
W^* = W_1^* \dot{\cup} \ldots \dot{\cup} W_k^*
$$

represents the normalization $W^* \xrightarrow{\nu} W$ of W (cf. [16]).

Let $S(W)$ denote the singular subset of W. Then for $1 \leqslant i \leqslant k$ $W_i \cap S(W)$ is a thin subset of W_i of codimension $\geqslant 1$.

Given on V a holomorphic function F of exponential type we can construct on W the gauge function g associated to F and for each i with

$1 \leqslant i \leqslant k$ we can consider on $W_i^* - \nu_i^{-1}(W_i \cap S(W))$ the function $g^{(i)} = g \circ \nu_i$. This is a plurisubharmonic function which is locally bounded in the neighborhood of every point $w^* \in W_i^*$. As W_i^* is normal the function $g^{(i)}$ extends in a unique way to a plurisubharmonic function (cf. [10] and [11])

$$g^{(i)}\colon W_i^* \to \mathbb{R} \cup \{-\infty\}\,.$$

The collection of the functions $g^{(i)}$ defines a plurisubharmonic function

$$g^*\colon W^* \to \mathbb{R} \cup \{-\infty\}$$

and we have, for every $z \in W$

$$g(z) = \sup \{g^*(\xi) \mid \xi \in W^*,\ \nu(\xi) = z\}\,.$$

Note that the plurisubharmonic function g^* is uniquely defined in terms of the gauge function g. The gauge function is plurisubharmonic on $W - S(W)$ but not plurisubharmonic on the whole variety W but it is uniquely defined in terms of the function g^*.

SECTION 11

The Classical Case

We mean by this the case in which $V = \mathbb{C}^n$ and therefore F is an entire function on $\mathbb{C}^n$ of exponential type. Set for every $\lambda > 0$

$$\varphi_\lambda(z) = \frac{\log|F(\lambda z)|}{\lambda} .$$

This is a plurisubharmonic function which is uniformly bounded from above on any compact set for any value of $\lambda > 1$.

The associated gauge function has been defined by

$$g(z) = \max \lim_{\substack{w \to z \\ \lambda \to +\infty}} \varphi_\lambda(w) .$$

The following proposition shows that this definition agrees with the classical one.

PROPOSITION 12. *Under the above assumptions we have*

$$g(z) = \max \lim_{w \to z} \max \lim_{\lambda \to \infty} \varphi_\lambda(w) .$$

PROOF. Let $\psi(z) = \max \lim\limits_{w \to z} \max \lim\limits_{\lambda \to \infty} \varphi_\lambda(w)$. Thus

$$g(z) = \inf_{\varepsilon > 0} \inf_{M > 1} \sup_{\substack{|w - z| < \varepsilon \\ w \neq z}} \sup_{\lambda > M} \varphi_\lambda(w)$$

$$\psi(z) = \inf_{\varepsilon > 0} \sup_{\substack{|w - z| < \varepsilon \\ w \neq z}} \inf_{M > 1} \sup_{\lambda > M} \varphi_\lambda(w).$$

For a function of two variables $G(x, y)$ defined for $x \in A$, $y \in B$ we have

$$\inf_{x \in A} \sup_{y \in B} G(x, y) \geqslant \sup_{y \in B} \inf_{x \in A} G(x, y).$$

Thus from this remark we derive that

$$g(z) \geqslant \psi(z) .$$

To prove the opposite inequality we proceed as follows.

Let $z_0 \in \mathbb{C}^n$ be fixed and let $\varepsilon > 0$. By the definition of ψ we can find a neighborhood $U(z_0)$ of z_0 in $\mathbb{C}^n$ such that

$$\max_{\lambda \to \infty} \lim \varphi_\lambda(w) < \psi(z_0) + \varepsilon \qquad \forall w \in U(z_0) .$$

We now apply the following lemma due to Hörmander (in the case of one variable but easily extended to the general case) (cf. [14] theorem 1. 6. 13 p. 21).

LEMMA. *Let $\{\varphi_\lambda\}_{\lambda > 0}$ be a family of plurisubharmonic functions defined in an open set U of $\mathbb{C}^n$. We assume that*

i) *on the open set U the functions φ_λ are uniformly bounded from above*

ii) *we have for some constant c and every $w \in U$*

$$\max_{\lambda \to \infty} \lim \varphi_\lambda(w) < c .$$

Then given $\sigma > 0$, given K compact in U there exists a $\lambda_0 = \lambda_0(\sigma, K) > 0$ such that

for $\lambda > \lambda_0$ and for $z \in K$ we have

$$\varphi_\lambda(z) < c + \sigma .$$

Let $U'(z_0)$ be an open neighborhood of z_0 contained and relatively compact in $U(z_0)$. By the above lemma we can find $\lambda_0 > 0$ such that

$$\forall w \in U'(z_0) , \qquad \forall \lambda > \lambda_0 , \qquad \varphi_\lambda(w) < \psi(z_0) + 2\varepsilon .$$

From this inequality we deduce that

$$g(z_0) \leqslant \psi(z_0) + 2\varepsilon .$$

Since $\varepsilon > 0$ is arbitrary, we must have therefore

$$g(z_0) \leqslant \psi(z_0) .$$

This concludes the proof.

SECTION 12

Sufficient Conditions for Analytic Convexity

We resume the notations of n. 8 and n. 9.

Let $A_0(D)$: $\mathcal{E}^{p_0}(\Omega) \to \mathcal{E}^{p_1}(\Omega)$ be a differential operator with constant coefficients defined on $\mathbb{R}^n$. For $\mathcal{S} = \mathbb{C}[\xi_1, \dots, \xi_n]$ we consider the corresponding $\mathcal{S}$-module

$$\operatorname{Im}\left\{\mathcal{S}^{p_1} \xrightarrow{{}^tA_0(\xi)} \mathcal{S}^{p_0}\right\}.$$

Of this module we consider a primary decomposition and we denote by $\mathfrak{p}_1, \dots, \mathfrak{p}_k$ the associated prime ideals. We set

$$V_i = \{\xi \in \mathbb{C}^n | p(\xi) = 0 \ \forall p \in \mathfrak{p}_i\}$$

for $1 \leqslant i \leqslant k$. For each algebraic variety V_i we consider the corresponding asymptotic variety $W = V_i^0$. We decompose W into irreducible components and we denote by

$$W_1, W_2, \dots, W_l$$

the distinct irreducible components that appear in an irreducible decomposition of at least one of the asymptotic varieties $V_1^0, \dots, V_k^0$. Let

$$W_1^* \xrightarrow{\nu_1} W_1, \dots, W_l^* \xrightarrow{\nu_l} W_l$$

be the normalizations of the varieties $W_1, \dots, W_l$ respectively. For a plurisubharmonic function

$$g^*: W_i^* \to \mathbb{R} \cup \{-\infty\}$$

we define

$$g: W_i \to \mathbb{R} \cup \{-\infty\}$$

by setting

$$g(z) = \sup \{g^*(\xi) \,|\nu_i(\xi) = z, \xi \in W_i^*\} .$$

Thus g is plurisubharmonic on $W_i - S(W_i)$ where $S(W_i)$ denotes the singular set of W_i.

We have the following

PROPOSITION 13. *Let ω be an open convex set in $\mathbb{R}^n$. The following is a sufficient condition for ω to be analytically convex with respect to the differential operator A_0 (i.e. that $H^1(\omega, \mathcal{A}_{A_0}) = 0$).*

For every compact convex set $K \subset \omega$ we can find $\delta > 0$ and a compact convex set K' with $K \subset K' \subset \omega$ such that:

for any j with $1 \leqslant j \leqslant l$ and any choice of a plurisubharmonic function $g^*: W_j^* \to \mathbb{R} \cup \{-\infty\}$ such that the following conditions are satisfied

$$(a) \qquad g(z) \leqslant H_K(z) + \delta|z| \qquad \forall z \in W_j$$

$$(b) \qquad g(z) \leqslant 0 \qquad \forall z \in W_j \cap i\mathbb{R}^n$$

then g satisfies also the condition

$$(c) \qquad g(z) \leqslant H_{K'}(z) \qquad \forall z \in W_j .$$

PROOF. α) Let us first remark the following. Let $\eta \in \mathcal{E}'(\mathbb{C}^n)$ be a distribution with compact support. We consider its Laplace transform

$$\tilde{\eta}(\xi) = \eta(\exp(\langle \xi, z \rangle))$$

where $\langle \xi, z \rangle = \sum_1^n \xi_i z_i$. Let $\mathfrak{L}(\xi, D_\xi)$ be a differential operator in $D_\xi = (\partial/\partial\xi_1, \ldots, \partial/\partial\xi_n)$ with coefficients polynomial in ξ.

Then the holomorphic function

$$\mathfrak{L}(\xi, D_\xi)\tilde{\eta}(\xi) = F(\xi)$$

is an entire function of exponential type.

Indeed $\tilde{\eta}(\xi)$ is an entire function thus $F(\xi)$ is also entire. We have

$$\begin{aligned} \mathfrak{L}(\xi, D_\xi)\tilde{\eta}(\xi) &= \eta(\mathfrak{L}(\xi, D_\xi) \exp(\langle \xi, z \rangle)) \\ &= \eta(\mathfrak{L}(\xi, z) \exp(\langle \xi, z \rangle)) . \end{aligned}$$

There exist a compact $K \subset \mathbb{C}^n$ and a constant $c > 0$ such that, for any holomorphic function f on $\mathbb{C}^n$ we have

$$|\eta(f)| \leqslant c\|f\|_K .$$

From this and the above expression we deduce that for some constant $c_1 > 0$ we have

$$|\mathfrak{L}(\xi, D_\xi)\tilde{\eta}(\xi)| \leqslant c_1 \exp(c_1|\xi|) .$$

β) Let now $\mu \in \mathcal{O}'_{A_0}(\mathbb{C}^n)$ be an A_0-analytic functional and let $K'' \subset \omega$ be any convex compact set.

We consider an extension η of μ to $\Gamma(\mathbb{C}^n, \mathcal{E}^{p_0})$ so that

$$\eta = \begin{pmatrix} \eta_1 \\ \vdots \\ \eta_{p_0} \end{pmatrix} \in \mathcal{E}'^{p_0}(\mathbb{C}^n)$$

and let $\tilde{\eta}(\xi)$ denote the Laplace transform of η.

For any j with $1 \leqslant j \leqslant k$ we consider the function $\mathfrak{L}_j(\xi, D_\xi)\tilde{\eta}(\xi) = F(\xi)$ where $\mathfrak{L}_j$ is the Palamodov operator associated to the variety V_j.

By the remark made in α) we have that each component of F is of exponential type. We denote by F_j any one of these components.

Now assume that μ is carried by K_δ and K''. This can be expressed in term of the gauge function g_j associated to F_j and defined on the asymptotic variety V_j^0 by virtue of proposition 9 and proposition 7. These conditions are

$$(\alpha) \qquad g_j(z) \leqslant H_K(z) + \delta|z| \qquad \forall z \in V_j^0$$

$$g_j(z) \leqslant H_{K''}(z) \qquad \forall z \in V_j^0$$

and for any j with $1 \leqslant j \leqslant k$. In particular since $K'' \subset \mathbb{R}^n$ we get from the second of these conditions $H_{K''}(z) = 0$ if $z \in i\mathbb{R}^n$ and thus

$$(\beta) \qquad g_j(z) \leqslant 0 \qquad \forall z \in V_j^0 \cap i\mathbb{R}^n$$

$1 \leqslant j \leqslant k$.

Let W_α be any one of the irreducible components of V_j^0, $1 \leqslant \alpha \leqslant s(j)$. Let g_α^* denote the plurisubharmonic function defined on W_α^*, the normalization of W_α, by $\nu_\alpha^* g_j$ on $W_\alpha^* - \nu_\alpha^{-1}$(singular set of V_j^0), and by the Riemann extension theorem. Set also, for $z \in W_\alpha$,

$$g_\alpha(z) = \sup\{g_\alpha^*(\xi) \mid \xi \in W_\alpha^*, \nu_\alpha(\xi) = z\} .$$

Then we have

$$g_\alpha(z) \leqslant g_j(z)$$

since $g_j(z) = \sup\limits_\alpha g_\alpha(z)$ for any $z \in V_j^0$. Since g_j verifies conditions (α) and (β) it follows that for $1 \leqslant \alpha \leqslant s(j)$ g_α verifies conditions (a) and (b) of the theorem. Thus by assumption

$$g_\alpha(z) \leqslant H_{K'}(z) \qquad \forall z \in W_\alpha$$

for $1 \leqslant \alpha \leqslant s(j)$. This implies that

$$(\gamma) \qquad g_j(z) \leqslant H_{K'}(z) \qquad \forall z \in V_j^0$$

as $g_j(z) = \sup g_\alpha(z)$ for $z \in V_j^0$. But this implies that any component F_j of $F = \mathfrak{L}_j(\xi, D_\xi)\tilde{\eta}(\xi)$ satisfies the condition:

$\forall \varepsilon > 0$ there exists a constant $c_\varepsilon > 0$ such that

$$|F_j(z)| \leqslant c_\varepsilon \exp\left(H_{K'}(z) + \varepsilon |z|\right) \qquad \forall z \in V_j\,.$$

This by virtue of proposition 9. But then by proposition 7 this brings as a consequence that μ is carried also by K'. Therefore the carrier condition (C) is satisfied and consequently $H^1(\omega, \mathcal{A}_{A_0}) = 0$ as we wanted to prove.

b) Let W denote an algebraic subvariety of the space $\mathbb{C}^n$ and let $S(W)$ denote the singular set of W.

A function $\varphi\colon W \to \mathbb{R} \cup \{-\infty\}$ will be called *weakly plurisubharmonic* on W if it satisfies the following conditions

i) for any $\Theta \in W$ we have

$$\varphi(\Theta) = \max \lim_{\substack{y \to \Theta \\ y \in W}} \varphi(y)$$

ii) φ is plurisubharmonic on $W - S(W)$.

In particular from i) it follows that φ is upper-semicontinuous and thus locally bounded from above and therefore, if W is a normal variety, it follows that φ is plurisubharmonic everywhere on W, because of condition i).

Let $W^* \xrightarrow{\nu} W$ be the normalization of W. Consider the function $\nu^*\varphi = \varphi \circ \nu$ on $W^* - \nu^{-1}(S(W))$; that function extends to the whole W^* as a plurisubharmonic function φ^* because of the analog of Riemann extension theorem (cf. [10]). Therefore for any point $\Theta^* \in W^*$ we will have in particular

$$\varphi^*(\Theta^*) = \max \lim_{\substack{y^* \to \Theta^* \\ y^* \in W^*}} \varphi^*(y^*) .$$

The relation between φ and φ^* is then given by the following definition of φ in terms of φ^*:

$\forall \Theta \in W$ we have

$$\varphi(\Theta) = \sup \{\varphi^*(\Theta^*) | \Theta^* \in W^*, \nu(\Theta^*) = \Theta\} .$$

Moreover we remark that if $\varphi^*\colon W^* \to \mathbb{R} \cup \{-\infty\}$ is any plurisubharmonic function defined on the normalization W^* of W then the function φ given by the above definition on W will be a weakly plorisubharmonic function on W.

c) We can now state the main theorem we want to prove. We resume the notations of point *a*) in this section.

THEOREM 3. *Let ω be an open and convex set in $\mathbb{R}^n$.*

A necessary and sufficient condition for the analytic convexity of ω i.e. in order that $H^1(\omega, \mathcal{A}_{A_0}) = 0$, is that the following statement holds.

For any compact set $K \subset \omega$ there exists a $\delta > 0$ and a convex compact set K' with $K \subset K' \subset \omega$ such that:

for any j with $1 \leqslant j \leqslant l$ and any choice of a weakly plurisubharmonic function $\varphi\colon W_i \to \mathbb{R} \cup \{-\infty\}$ which verifies the following conditions

(*a*) $$\varphi(z) \leqslant H_K(z) + \delta|z| \qquad \forall z \in W_j$$

(*b*) $$\varphi(z) \leqslant 0 \qquad \forall z \in W_j \cap i\mathbb{R}^n$$

then φ also satisfies the condition

(*c*) $$\varphi(z) \leqslant H_{K'}(z) \qquad \forall z \in W_j .$$

Because of the above remarks on weakly plurisubhamonic functions and because of proposition 13 the condition of theorem 3 has already been proved sufficient.

To establish the necessity of the same condition we will transform it successively into equivalent forms till we obtain a condition on particular entire functions that we will prove to be also a necessary condition for analytic convexity. This will be the content of the following sections.

SECTION 13

Reduction to Primary Modules

We resume the notations of section 8. Let $A_0(D)\colon \mathcal{E}^{p_0}(\Omega) \to \mathcal{E}^{p_1}(\Omega)$ be a differential operator with constant coefficients given in $\mathbb{R}^n$. We consider the $\mathcal{S}$-module ($\mathcal{S} = \mathbb{C}[\xi_1, \ldots, \xi_n]$)

$$N = \operatorname{Im}\left\{\mathcal{S}^{p_1} \xrightarrow{{}^tA_0(\xi)} \mathcal{S}^{p_0}\right\}$$

and a primary decomposition of $N = \bigcap_{i=1}^{k} Q(\mathfrak{p}_i)$ where $\mathfrak{p}_i$ are the prime ideals of $\operatorname{Ass}(\mathcal{S}^{p_0}/N)$. For each module $Q(\mathfrak{p}_i)$ we consider a presentation

$$Q(\mathfrak{p}_i) = \operatorname{Im}\left\{\mathcal{S}^{q_i} \xrightarrow{{}^tA_i(\xi)} \mathcal{S}^{p_0}\right\}$$

and correspondingly we consider the differential operator

$$A_i(D)\colon \mathcal{E}^{p_0}(\Omega) \to \mathcal{E}^{q_i}(\Omega)$$

with constant coefficients. There exist differential operators with constant coefficients $\beta_i(D)$ such that we have identically for every i with $1 \leqslant i \leqslant k$

$$A_0(D) = \beta_i(D) A_i(D) .$$

We denote by $\mathcal{A}_{A_i}$ the sheaf of germs of complex valued real analytic solution $u \in \mathcal{A}^{p_0}$ of the equation $A_i(D)u = 0$ for $0 \leqslant i \leqslant k$.

PROPOSITION 14. *Let ω be open and convex in $\mathbb{R}^n$. If we have*

$$H^1(\omega, \mathcal{A}_{A_i}) = 0 \qquad \textit{for } 1 \leqslant i \leqslant k$$

then we have also

$$H^1(\omega, \mathcal{A}_{A_0}) = 0 .$$

PROOF. Set

$$M_i = \mathcal{P}^{p_0}/Q(\mathfrak{p}_i) \qquad \text{for } 1 \leqslant i \leqslant k \text{ and } M = \mathcal{P}^{p_0}/N\,.$$

Since $N = \bigcap_{i=1}^{k} Q(\mathfrak{p}_i) \subset Q(\mathfrak{p}_i)$ we have natural maps

$$\nu_i \colon M \to M_i$$

and thus a natural map

$$\nu \colon M \to M_1 \oplus \ldots \oplus M_k$$

defined by $\nu(m) = \Sigma\nu_i(m)$ for $m \in M$. This map, since $N = \bigcap_{i=1}^{k} Q(\mathfrak{p}_i)$, is injective and thus we can write an exact sequence

$$(*) \qquad 0 \to M \to M_1 \oplus \ldots \oplus M_k \to L \to 0$$

where L is defined by the exactness of the sequence. By its definition L is a $\mathcal{P}$-module of finite type and we can consider a presentation

$$\mathcal{P}^q \xrightarrow{{}^tE(\xi)} \bigoplus_1^k \mathcal{P}^{p_0} \to L \to 0$$

of the module L as coker of a $\mathcal{P}$-homomorphism between free $\mathcal{P}$-modules of finite type.

Now the $\mathcal{P}$-module $\mathcal{A}_x$ of germs of complex valued real analytic functions is an injective module (for $p \in \mathcal{P}$, $\alpha \in \mathcal{A}_x$ we define $p \circ \alpha = p(D)\alpha$) (cf. [4] proposition 1). Applying the functor $\mathrm{Hom}_{\mathcal{P}}(\cdot, \mathcal{A}_x)$ to the sequence $(*)$ we obtain an exact sequence

$$(**) \qquad 0 \to \mathcal{A}_E \to \mathcal{A}_{A_1} \oplus \ldots \oplus \mathcal{A}_{A_k} \xrightarrow{\alpha} \mathcal{A}_{A_0} \to 0$$

which gives an exact sequence of sheaves. Here $\mathcal{A}_E$ denotes the sheaf of germs $u = \begin{pmatrix} u_1 \\ \vdots \\ u_k \end{pmatrix}$ with $u_j \in \mathcal{A}^{p_0}$ such that $E(D)u = 0$ and α is the natural map given by $\alpha(u_1 \oplus \ldots \oplus u_k) = \Sigma u_i$. These are properties that one verifies directly from the type of the maps given in $(*)$. From the exact sequence of sheaves $(**)$ one deduces for ω open in $\mathbb{R}^n$ an exact cohomology sequence

$$\bigoplus_1^k H^1(\omega, \mathcal{A}_{A_i}) \xrightarrow{\alpha^*} H^1(\omega, \mathcal{A}_{A_0}) \to H^2(\omega, \mathcal{A}_E)\,.$$

Since ω is open and convex we have $H^2(\omega, \mathcal{A}_E) = 0$ (cf. [4] theorem 4). This implies the surjectivity of the map α^* and thus the conclusion of our proposition.

We consider now the $\mathcal{P}$-module

$$N_i = \Big\{\bigcap_{j\neq i} Q(\mathfrak{p}_j)\Big\}/N \qquad 1 \leqslant i \leqslant k$$

and a presentation of this module as the cokernel of a $\mathcal{P}$-homomorphism between free modules of finite type. We have thus an exact sequence

$$\mathcal{P}^{k_i} \xrightarrow{{}^tB_i(\xi)} \mathcal{P}^{h_i} \to N_i \to 0\,.$$

We claim that

$$\mathcal{B}_i = \operatorname{Im}\Big\{\mathcal{P}^{k_i} \xrightarrow{{}^tB_i(\xi)} \mathcal{P}^{h_i}\Big\}$$

is a primary submodule of $\mathcal{P}^{h_i}$ with associated prime $\mathfrak{p}_i$ $1 \leqslant i \leqslant k$. To this end we show that for $\forall m \in N_i - \{0\}$

$$\operatorname{Ann}(m) \subset \mathfrak{p}_i$$

and that on the other hand we have

$$\mathfrak{p}_i \subset \sqrt{\operatorname{Ann}(N_i)}\,.$$

Indeed if $p \in \mathcal{P}$ and $pm = 0$ there exists $\mathfrak{X} \in \bigcap_{j\neq i} Q(\mathfrak{p}_j) - N$ (whose image is m) such that $p\mathfrak{X} \in N$. But $\mathfrak{X} \notin Q(\mathfrak{p}_i)$, thus $p \in \mathfrak{p}_i$. Hence $\operatorname{Ann}(m) \subset \mathfrak{p}_i$.

Conversely $\forall p \in \mathfrak{p}_i$ we have $p^k \mathcal{P}^{p_0} \subset Q(\mathfrak{p}_i)$ for some integer $k > 0$. Thus $p^k N_i = 0$ and therefore $p \in \sqrt{\operatorname{Ann}(N_i)}$. Hence $\mathfrak{p}_i \subset \sqrt{\operatorname{Ann}(N_i)}$. We explicitly remark that the characteristic variety associated to the operator B_i is the variety

$$V_i = \{\xi \in \mathbb{C}^n \,|\, p(\xi) = 0 \;\forall p \in \mathfrak{p}_i\}$$

which we had already encountered as the characteristic variety of the operator A_i for $1 \leqslant i \leqslant k$.

We denote as usual by $\mathcal{A}_{B_i}$ the sheaf of germs of complex valued real analytic solutions $u \in \mathcal{A}^{h_i}$ of the equation $B_i(D)u = 0$ for $1 \leqslant i \leqslant k$.

Proposition 15. *Let ω be an open convex set in $\mathbb{R}^n$. If we have*

$$H^1(\omega, \mathcal{A}_{A_0}) = 0$$

then we also have

$$H^1(\omega, \mathcal{A}_{B_i}) = 0 \quad \text{for } 1 \leqslant i \leqslant k\,.$$

PROOF. We have natural exact sequences of $\mathcal{P}$-modules

$$(*) \qquad 0 \to N_i \to M \to S_i \to 0$$

where $S_i = \mathcal{P}^{p_0} / \bigcap_{j \neq i} Q(\mathfrak{p}_j)$. We define ${}^tD_i(\xi)$ by

$$\bigcap_{j \neq i} Q(\mathfrak{p}_j) = \operatorname{Im}\left\{\mathcal{P}^{l_i} \xrightarrow{{}^tD_i(\xi)} \mathcal{P}^{p_0}\right\}.$$

Applying to $(*)$ the functor $\operatorname{Hom}_{\mathcal{P}}(\cdot, \mathcal{A}_x)$ we obtain an exact sequence

$$0 \to \mathcal{A}_{D_i} \to \mathcal{A}_{A_0} \to \mathcal{A}_{B_i} \to 0$$

which is also an exact sequence of sheaves. Here $\mathcal{A}_{D_i}$ is the sheaf of germs of complex valued real analytic function $u \in \mathcal{A}^{p_0}$ such that $D_i(D)u = 0$. This because of the injectivity of $\mathcal{A}_x$ as a $\mathcal{P}$-module (cf. the previous proof).

From the exact cohomology sequence we derive an exact sequence for every i with $1 \leqslant i \leqslant k$

$$H^1(\omega, \mathcal{A}_{A_0}) \to H^1(\omega, \mathcal{A}_{B_i}) \to H^2(\omega, \mathcal{A}_{D_i})\,.$$

Because ω is convex open we have $H^2(\omega, \mathcal{A}_{D_i}) = 0$ and thus a surjective map

$$H^1(\omega, \mathcal{A}_{A_0}) \to H^1(\omega, \mathcal{A}_{B_i})$$

for every i. This implies the statement of the proposition.

We have already proved the sufficiency of the conditions stated in theorem 3 for the analytic convexity of a convex open set ω in general for a differential operator A_0. It would have been enough to prove the sufficiency of the same conditions for differential operators associated to primary $\mathcal{P}$-modules like the operators A_i, $1 \leqslant i \leqslant k$, since, by proposition 14 we would have recovered the sufficiency of the corresponding conditions for the operator A_0 itself. Conversely it is enough to prove the necessity of the conditions of theorem 3 for differential operators associated to primary $\mathcal{P}$-modules, like the operators B_i, $1 \leqslant i \leqslant k$. In fact, by proposition 15, we deduce then the necessity of the conditions stated in theorem 3 for a general differential operator A_0.

Therefore there is no loss of generality in assuming for the proof of the necessity of the conditions of theorem 3, that we deal with differential operators associated to primary $\mathcal{P}$-modules.

SECTION 14

Equivalent Formulations of Phragmén-Lindelöf Principle

a) By W we denote an algebraic cone in $\mathbb{C}^n$ i.e. an affine algebraic subvariety of $\mathbb{C}^n$ with the property

$$\text{if} \quad w \in W\,, \quad \lambda \in \mathbb{C} \quad \text{then } \lambda w \in W\,.$$

Let $K \subset K'$ be compact convex sets in $\mathbb{R}^n$ and let us assume that K' is a neighborhood of K (i.e. $K \subset \mathring{K}'$).

The following statement will be called the *Phragmén-Lindelöf principle on W*:

(I) « *There is a $\delta > 0$ such that for every weakly plurisubharmonic function φ on W which satisfies the conditions*

$$(a) \qquad \varphi(\xi) \leqslant H_K(\xi) + \delta|\xi| \qquad \forall \xi \in W$$

$$(b) \qquad \varphi(\xi) \leqslant 0 \qquad \forall \xi \in W \cap i\mathbb{R}^n$$

one has also

$$(c) \qquad \varphi(\xi) \leqslant H_{K'}(\xi) \qquad \forall \xi \in W \text{ »}.$$

The purpose of the present section is to give to the Phragmén-Lindelöf principle on W a sequence of equivalent formulations.

b) For $\Theta \in \mathbb{C}^n$ and $r > 0$ we set

$$B(\Theta, r) = \{\xi \in \mathbb{C}^n \mid |\xi - \Theta| < r\}$$

where as usual $|\cdot|$ denotes the euclidean norm.

We first establish the following

Lemma 11. *Let K be a convex compact subset of $\mathbb{R}^n$, let $\delta > 0$ and let $0 < r < 1$ be given.*

We can find $\sigma > 0$ such that

for any $\Theta \in W$ with $|\operatorname{Im}\Theta| = 1$, $|\operatorname{Re}\Theta| < r/2$

for any weakly plurisubharmonic function $\varphi\colon W \cap B(\Theta, r) \to \mathbb{R} \cup \{-\infty\}$ which satisfies the conditions

$$\begin{aligned}&\varphi(\xi) \leqslant H_K(\xi) + \sigma|\xi| && \forall \xi \in W \cap B(\Theta, r)\\ &\varphi(\xi) \leqslant 0 && \forall \xi \in W \cap i\mathbb{R}^n \cap B(\Theta, r)\end{aligned}$$

we can select a weakly plurisubharmonic function $\psi\colon W \to \mathbb{R} \cup \{-\infty\}$ satisfying the conditions

$$\begin{aligned}&\psi(\xi) \leqslant H_K(\xi) + \delta|\xi| && \forall \xi \in W\\ &\psi(\xi) \leqslant 0 && \forall \xi \in W \cap i\mathbb{R}^n\\ &\psi(\Theta) \geqslant \varphi(\Theta)\,.\end{aligned}$$

Proof. α) Let $\chi \in \mathcal{D}(\mathbb{R}^n)$ be a real valued C^∞ function with compact support in $\mathbb{R}^n$. We assume that

$$\begin{aligned}&\chi \geqslant 0 && \text{on } \mathbb{R}^n\\ &\chi(x) = \chi(-x) && \text{for all } x \in \mathbb{R}^n\\ &\int \chi\, dx = 1 && (\text{where } dx \text{ is the Lebesgue measure in } \mathbb{R}^n)\\ &\operatorname{supp}\chi \subset \{x \in \mathbb{R}^n |\ |x| < 1\}\,.\end{aligned}$$

We consider for $\xi \in \mathbb{C}^n$

$$\tilde{\chi}(\xi) = \int_{\mathbb{R}^n} \chi(x) \exp(\langle \xi, x\rangle)\, dx$$

which is the Laplace transform of χ. This is an entire function on $\mathbb{C}^n$ and therefore

$$\phi(\xi) = \log|\tilde{\chi}(\xi)|$$

is plurisubharmonic on $\mathbb{C}^n$. We claim that

$$(*) \qquad \phi(\xi) < |\operatorname{Re}\xi| \qquad \forall \xi \in \mathbb{C}^n - \{0\}\,.$$

Indeed let $M = \sup\{|x| \ |x \in \operatorname{supp}\chi\}$. We have $M < 1$ and therefore, if $\operatorname{Re}\xi \neq 0$, we do have

$$|\tilde{\chi}(\xi)| \leqslant \int_{\mathbf{R}^n} \chi(x) \exp(\langle \operatorname{Re}\xi, x\rangle)\,dx \leqslant \sup_{x\in\operatorname{supp}\chi} \exp(\langle \operatorname{Re}\xi, x\rangle) \int_{\mathbf{R}^n} \chi(x)\,dx$$
$$\leqslant \exp(M|\operatorname{Re}\xi|) < \exp(|\operatorname{Re}\xi|)\,.$$

At a point $\zeta \neq 0$ where $\operatorname{Re}\zeta = 0$, i.e. where $\zeta = i\xi$ with $\xi \in \mathbf{R}^n - \{0\}$ we have

$$\tilde{\chi}(i\xi) = \int \chi(x) \cos\langle \xi, x\rangle\,dx \qquad \text{(since } \chi \text{ is even)}$$
$$< \int_{\mathbf{R}^n} \chi(x)\,dx = 1$$

because $1 - \cos\langle \xi, x\rangle > 0$ on some non empty open subset of $\operatorname{supp}\chi$. The function $\varphi(\xi) - |\operatorname{Re}\xi|$ is upper semicontinuous and on a circular shell $r_1 < |\xi| < r_2$, with $r_2 > r_1 > 0$, has a negative maximum $-\mu(r_1, r_2) < 0$. Thus we have

$$(**) \qquad \phi(\xi) < |\operatorname{Re}\xi| - \mu \qquad \text{for } r_1 < |\xi| < r_2$$

with $\mu > 0$ and for $r_2 > r_1 > 0$.

Finally we remark that

$$(***) \qquad \phi(\xi) \geqslant 0 \qquad \text{for } \xi \in \mathbf{R}^n\,.$$

Indeed for $\xi \in \mathbf{R}^n$, since χ is an even function, we have

$$\tilde{\chi}(\xi) = \int_{\mathbf{R}^n} \chi(x) \frac{\exp(\langle \xi, x\rangle) + \exp(-\langle \xi, x\rangle)}{2}\,dx \geqslant \int_{\mathbf{R}^n} \chi(x)\,dx = 1$$

since, for $t \in \mathbf{R}$, $\cos ht = (e^t + e^{-t})/2 \geqslant 1$.

β) We set

$$\eta = i \operatorname{Im}\Theta$$

for a fixed Θ chosen as in the statement of the lemma. We consider the weakly plurisubharmonic function on $W \cap B(\Theta, r)$

$$\psi_1(\xi) = \varphi(\xi) + \frac{\delta}{2}\phi(\xi - \eta) \qquad \text{for } \xi \in W \cap B(\Theta, r)\,.$$

If we have choosen σ with $0 < \sigma \leqslant \delta/2$ we have

$$(****) \qquad \psi_1(\xi) \leqslant H_K(\xi) + \sigma|\xi| + (\delta/2)|\mathrm{Re}\,\xi|$$
$$\leqslant H_K(\xi) + \delta|\xi| \qquad \text{for } \xi \in W \cap B(\Theta, r)\,.$$

On the boundary of $B(\Theta, r)$ we have

$$|\xi - \eta| \leqslant |\xi - \Theta| + |\Theta - \eta|$$
$$\leqslant r + \frac{r}{2} = \frac{3}{2}r$$

and

$$|\xi - \eta| \geqslant |\xi - \Theta| - |\Theta - \eta|$$
$$\geqslant r - \frac{r}{2} = \frac{r}{2}\,.$$

We apply inequality (**) for $r_1 = r/2$ and $r_2 = (\frac{3}{2})r$, so that with $\mu > 0$ depending only on r we have

$$\phi(\xi - \eta) < |\mathrm{Re}\,\xi| - \mu \qquad \text{for } \xi \in \partial B(\Theta, r)\,.$$

Note that on $B(\Theta, r)$

$$|\xi| \leqslant |\xi - \Theta| + |\Theta| \leqslant r + \frac{r}{2} + 1 < 3\,.$$

Therefore for ξ near the boundary of $W \cap B(\Theta, r)$ we have

$$\psi_1(\xi) = \varphi(\xi) + \frac{\delta}{2}\phi(\xi - \eta)$$
$$< H_K(\xi) + \sigma|\xi| + \frac{\delta}{2}|\mathrm{Re}\,\xi| - \frac{\delta}{2}\mu$$
$$< H_K(\xi) + \frac{\delta}{2}|\mathrm{Re}\,\xi| \qquad \text{provided } 0 < \sigma < \frac{\delta}{6}\mu\,.$$

γ) We now define the function $\psi\colon W \to \mathbb{R} \cup \{-\infty\}$ by setting

$$\psi(\xi) = \begin{cases} \max\left\{\psi_1(\xi), H_K(\xi) + \dfrac{\delta}{2}|\mathrm{Re}\,\xi|\right\} & \text{for } \xi \in W \cap B(\Theta, r) \\[2ex] H_K(\xi) + \dfrac{\delta}{2}|\mathrm{Re}\,\xi| & \text{for } \xi \in W - W \cap B(\Theta, r)\,. \end{cases}$$

This function is locally at any point of W the supremum of a weakly plurisubharmonic function and a plurisubharmonic function.

Therefore ψ is a weakly plurisubharmonic function.

This function satisfies the condition

$$\psi(\xi) \leqslant H_K(\xi) + \delta|\xi| \qquad \forall \xi \in W$$

because of (∗∗∗∗) and because of its definition.

Also for $\xi \in W \cap i\mathbb{R}^n$ we have $\phi(\xi - \eta) \leqslant 0$ because of (∗) and the fact that ϕ is plurisubharmonic on $\mathbb{C}^n$. Thus for $\xi \in W \cap i\mathbb{R}^n$ we have $\psi_1(\xi) \leqslant \leqslant \varphi(\xi)$ and therefore

$$\psi(\xi) \leqslant 0 \qquad \text{for } \xi \in W \cap i\mathbb{R}^n .$$

Finally for $\xi = \Theta$ we have $\phi(\Theta - \eta) = \phi(\operatorname{Re} \Theta) \geqslant 0$ because of (∗∗∗). Therefore

$$\psi(\Theta) \geqslant \psi_1(\Theta) \geqslant \varphi(\Theta) .$$

This achieves the proof.

As a first corollary of this lemma we deduce the following

PROPOSITION 16. *Let $K \subset K'$ be compact convex sets in $\mathbb{R}^n$ and let $K \subset \mathring{K}'$. Let r with $0 < r < 1$ be given.*

Then the Phragmén-Lindelöf principle (I) *on W is equivalent to the following statement*

(II) « *There is a $\sigma > 0$ such that for any choice of $\Theta \in W$ with $|\operatorname{Im} \Theta| = 1$, $|\operatorname{Re} \Theta| < r/2$ and for any choice of a weakly plurisubharmonic function φ on $W \cap B(\Theta, r)$ such that*

$$(\alpha) \qquad \varphi(\xi) \leqslant H_K(\xi) + \sigma|\xi| \qquad \forall \xi \in W \cap B(\Theta, r)$$

$$(\beta) \qquad \varphi(\xi) \leqslant 0 \qquad \forall \xi \in W \cap i\mathbb{R}^n \cap B(\Theta, r)$$

one has also that

$$(\gamma) \qquad \varphi(\Theta) \leqslant H_{K'}(\Theta) \text{ » .}$$

PROOF. (I) ⇒ (II). Let $\delta > 0$ be so chosen that I holds i.e. for any weakly plurisubharmonic function ψ on W which verifies

$$(a) \qquad \psi(\xi) \leqslant H_K(\xi) + \delta|\xi| \qquad \forall \xi \in W$$

$$(b) \qquad \psi(\xi) \leqslant 0 \qquad \forall \xi \in W \cap i\mathbb{R}^n$$

we also have

$$(c) \qquad \psi(\xi) \leqslant H_{K'}(\xi) \qquad \forall \xi \in W .$$

By lemma 11 we then can find $\sigma > 0$ such that for any $\Theta \in W$ with $|\mathrm{Im}\,\Theta| = 1$, $|\mathrm{Re}\,\Theta| < r/2$ and any choice of a weakly plurisubharmonic function φ on $W \cap B(\Theta, r)$ which satisfies (α) and (β) we can select a ψ in the previous class with $\psi(\Theta) \geqslant \varphi(\Theta)$.

Since $\psi(\Theta) \leqslant H_{K'}(\Theta)$ by assumption we must have also condition (γ) i.e. $\varphi(\Theta) \leqslant H_{K'}(\Theta)$.

(II) $\Rightarrow$ (I). Since $K \subset \mathring{K}'$, $\mathrm{dist}\,(K, C(K')) = \varepsilon > 0$. Now for $K_\varepsilon \cap \mathbb{R}^n$ we have

$$H_{K_\varepsilon \cap \mathbb{R}^n}(\xi) = H_K(\xi) + \varepsilon|\mathrm{Re}\,\xi| ,$$

thus we obtain

$$H_K(\xi) + \varepsilon|\mathrm{Re}\,\xi| \leqslant H_{K'}(\xi) \qquad \forall \xi \in \mathbb{C}^n$$

since $K_\varepsilon \cap \mathbb{R}^n \subset K'$.

Let $\sigma > 0$ be so chosen that statement II holds, and let us choose $\delta = \min\{\sigma, \varepsilon(1 + 2/r)^{-1}\}$. We claim that with that choice of δ statement I also holds. Let ψ be a weakly plurisubharmonic function on W for which conditions (a) and (b) are verified. We claim that condition (c) also holds for ψ.

Let $\xi \in W$ and assume first that $|\mathrm{Re}\,\xi| \geqslant (r/2)|\mathrm{Im}\,\xi|$. Then from condition ($a$) we derive,

$$\begin{aligned} \psi(\xi) &\leqslant H_K(\xi) + \delta|\xi| \leqslant H_K(\xi) + \delta|\mathrm{Re}\,\xi| + \delta|\mathrm{Im}\,\xi| \\ &\leqslant H_K(\xi) + \delta(1 + 2/r)|\mathrm{Re}\,\xi| \leqslant H_K(\xi) + \varepsilon|\mathrm{Re}\,\xi| \\ &\leqslant H_{K'}(\xi) \end{aligned}$$

therefore in this case condition (c) holds.

Assume now that $\xi \in W$ and $|\mathrm{Re}\,\xi| < (r/2)|\mathrm{Im}\,\xi|$. Then $|\mathrm{Im}\,\xi| \neq 0$ and

$$\Theta = \frac{\xi}{|\mathrm{Im}\,\xi|}$$

has the properties

$$\Theta \in W , \qquad |\mathrm{Im}\,\Theta| = 1 , \qquad |\mathrm{Re}\,\Theta| < \frac{r}{2} .$$

Consider the function

$$\varphi(\eta) = |\mathrm{Im}\,\xi|^{-1}\psi(\eta|\mathrm{Im}\,\xi|);$$

for ξ given, as a function of η, is weakly plurisubharmonic on $W \cap B(\Theta, r)$.

Moreover from (*a*) and (*b*) we obtain

$$\varphi(\eta) \leqslant H_K(\eta) + \delta|\eta| \leqslant H_K(\eta) + \sigma|\eta| \quad \text{for } \eta \in W \cap B(\Theta, r)$$
$$\varphi(\eta) \leqslant 0 \quad \text{for } \eta \in W \cap i\mathbb{R}^n \cap B(\Theta, r)\,.$$

Thus conditions (α) and (β) are satisfied. Therefore we must have also condition (γ) i.e. $\varphi(\Theta) \leqslant H_{K'}(\Theta)$. This says that

$$|\mathrm{Im}\,\xi|^{-1}\psi(\xi) \leqslant H_{K'}(|\mathrm{Im}\,\xi|^{-1}\xi) = |\mathrm{Im}\,\xi|^{-1}H_{K'}(\xi)\,.$$

Thus

$$\psi(\xi) \leqslant H_{K'}(\xi)$$

also in this case, and condition (*c*) holds for any $\xi \in W$.

c) Let us consider the decomposition of W into irreducible components

$$W = W_1 \cup W_2 \cup \ldots \cup W_l$$

where each W_j is an irreducible cone and not contained in $\bigcup_{s \neq j} W_s$, $1 \leqslant j \leqslant l$.

For an algebraic cone W denote by $\mathrm{I}(W, \delta)$ the statement of the Phragmén-Lindelöf principle for W with the constant $\delta > 0$. From the previous proposition we deduce that $\mathrm{I}(W, \delta)$ is equivalent to the union of the statements $\mathrm{I}(W_1, \sigma) \cup \ldots \cup \mathrm{I}(W_l, \sigma)$ for some $\sigma > 0$. More precisely we have the following

COROLLARY 1. *Let $K \subset K'$ be compact convex sets in $\mathbb{R}^n$ with $K \subset \mathring{K}'$. The Phragmén-Lindelöf principle* I *for W is equivalent to the following statement*

I$'$) « *There is a $\sigma > 0$ such that for any j with $1 \leqslant j \leqslant l$ and any weakly plurisubharmonic function φ on W_j which verifies*

(a_j) $$\varphi(\xi) \leqslant H_K(\xi) + \sigma|\xi| \quad \forall \xi \in W_j$$

(b_j) $$\varphi(\xi) \leqslant 0 \quad \forall \xi \in W_j \cap i\mathbb{R}^n$$

we also have

(c_j) $$\varphi(\xi) \leqslant H_{K'}(\xi) \quad \forall \xi \in W_j\,».$$

PROOF. We have $I(W_1, \sigma) \cup ... \cup I(W_l, \sigma) \Rightarrow I(W, \sigma)$. Thus (I') implies (I) with $\delta = \sigma$.

Conversely we have to prove that (I) implies (I'). Let $\varepsilon > 0$ be such that

$$H_K(\xi) + \varepsilon|\operatorname{Re}\xi| \leqslant H_{K'}(\xi) \qquad \forall \xi \in \mathbb{C}^n .$$

Suppose that $II(W, \delta)$ holds (statement (II) with δ in place of σ) and fix r with $0 < r < 1$. Set $\sigma = \min\{\delta, \varepsilon(1 + 2/r)^{-1}\}$. We want to show that $I(W_j, \sigma)$ holds for any j with $1 \leqslant j \leqslant l$.

Let $\xi \in W_j$ and assume first that $|\operatorname{Re}\xi| \geqslant (r/2)|\operatorname{Im}\xi|$. If φ is weakly plurisubharmonic on W_j and satisfies (a_j) we have

$$\begin{aligned}\varphi(\xi) &\leqslant H_K(\xi) + \sigma|\xi| \leqslant H_K(\xi) + \sigma|\operatorname{Re}\xi| + \sigma|\operatorname{Im}\xi| \\ &\leqslant H_K(\xi) + \varepsilon|\operatorname{Re}\xi| \\ &\leqslant H_{K'}(\xi) .\end{aligned}$$

Therefore there is nothing to prove in this case.

Let now $\xi \in W_j$ and assume that $|\operatorname{Re}\xi| < (r/2)|\operatorname{Im}\xi|$ so that for $\Theta = |\operatorname{Im}\xi|^{-1}\xi$ we have $|\operatorname{Im}\Theta| = 1$ and $|\operatorname{Re}\Theta| < r/2$.

Write $W = W_j \cup \bigcup_{s \neq j} W_s$. We can find a polynomial q_j homogeneous in its variables in $\mathbb{C}^n$ such that

$$q_j \Big| \bigcup_{s \neq j} W_s = 0 ; \qquad q_j \not\equiv 0 \text{ on } W_j .$$

We make the additional assumption $q_j(\xi) \neq 0$.

Let now φ be a weakly plurisubharmonic function on W_j satisfying conditions (a_j) and (b_j). For any $t > 0$ we consider the function

$$\psi_t(\eta) = \begin{cases} |\operatorname{Im}\xi|^{-1}\varphi(\eta|\operatorname{Im}\xi|) + t\log|q_j(\eta)| & \text{for } \eta \in W_j \cap B(\Theta, r) \\ -\infty \quad \text{for } \eta \in W \cap B(\Theta, r) - W_j \cap B(\Theta, r) . \end{cases}$$

Then $\psi_t(\eta)$ is a weakly plurisubharmonic function on $W \cap B(\Theta, r)$. Moreover it is not restrictive to assume that

$$|q_j(\eta)| \leqslant 1 \qquad \text{for } |\eta| \leqslant 3$$

so that we deduce that $\psi_t(\eta)$ satisfies the conditions

$$\begin{aligned}&\psi_t(\eta) \leqslant H_K(\eta) + \delta|\eta| \qquad \forall \eta \in W \cap B(\Theta, r) \\ &\psi_t(\eta) \leqslant 0 \qquad \forall \eta \in W \cap i\mathbb{R}^n \cap B(\Theta, r) .\end{aligned}$$

Consequently we have

$$|\mathrm{Im}\,\xi|^{-1}\varphi(\xi) + t\log|q_j(|\mathrm{Im}\,\xi|^{-1}\xi)| \leqslant |\mathrm{Im}\,\xi|^{-1}H_{K'}(\xi)\,.$$

Since we have assumed q_j homogeneous and $q(\xi)\neq 0$ letting t tend to zero for positive t's we obtain

$$\varphi(\xi)\leqslant H_{K'}(\xi)\,.$$

We have to remove the restriction $q_j(\xi)\neq 0$.

Let $W_j^* \xrightarrow{\nu_j} W_j$ be the normalization of W_j and let φ^* be the plurisubharmonic function on W_j^* defined as $\nu_j^*\varphi$ on $W_j^* - \nu_j^{-1}$ (singular set of W_j) and by Riemann extension theorem.

Let $A = \{w^* \in W_j^* | q_j(\nu_j(w^*)) = 0\}$. Then A is a thin subset of W_j^* and on any point $a\in A$ we have by the quoted Riemann theorem

$$\varphi^*(a) = \max \lim_{\substack{w^*\to a\\ w^*\in W^*-A}} \varphi^*(w^*)\,.$$

This shows that since on $W^* - A$ we have

$$\varphi^*(w^*)\leqslant H_{K'}(\nu_j(w^*))$$

the same inequality holds on the whole of W^*, because $H_{K'}(\nu_j(w^*))$ is a continuous function. But for $\xi\in W$ we have

$$\varphi(\xi) = \sup\{\varphi^*(w^*)|w^*\in W^*,\ \nu(w^*) = \xi\}\,.$$

This shows that we have $\varphi(\xi)\leqslant H_{K'}(\xi)$ for any $\xi\in W$.

COROLLARY 2. *Let $K\subset K'$ be compact convex sets in $\mathbb{R}^n$ with $K\subset \mathring{K}'$. The Phragmén-Lindelöf principle* I *for W is equivalent to the following statement*

(I″) « *There is a $\sigma>0$ such that for every j with $1\leqslant j\leqslant l$ and for every weakly plurisubharmonic function φ on W_j which verifies*

(a_j') $\qquad \varphi(\xi)\leqslant H_K(\xi) + \sigma|\xi| \qquad \forall\xi\in W_j - S(W_j)$

(b_j') $\qquad \varphi(\xi)\leqslant 0 \qquad \forall\xi\in W_j\cap i\mathbb{R}^n - W_j\cap i\mathbb{R}^n\cap S(W_j)$

(*where $S(W_j)$ denotes the singular set of W_j*) *we also have*

(c_j') $\qquad \varphi(\xi)\leqslant H_{K'}(\xi) \qquad \forall\xi\in W_j$ ».

PROOF. If (I″) holds then, a fortiori, (I′) holds and thus (I) holds. Therefore (I″) ⇒ (I).

Conversely let us assume that (I) holds. Then (I′) holds, thus, for fixed r with $0 < r < 1$, II(W_j, δ) (statement II for W_j and δ in place of W and σ) for some $\delta > 0$ and all j with $1 \leqslant j \leqslant l$, also holds. We want to show that (I″) holds.

Let $\varepsilon > 0$ be such that

$$H_K(\xi) + \varepsilon|\mathrm{Re}\,\xi| \leqslant H_{K'}(\xi) \qquad \forall \xi \in \mathbb{C}^n$$

and set $\sigma = \min\{\delta, \varepsilon(1 + 2/r)^{-1}\}$. Let φ be a weakly plurisubharmonic function on W_j which verifies (a_j') and (b_j').

We first claim that we have

$$\varphi(\xi) \leqslant H_K(\xi) + \sigma|\xi| \qquad \forall \xi \in W_j\,.$$

This is proved by lifting φ to the normalization W_j^* of W_j as we have done at the end of the proof of corollary 1.

Let now $\xi \in W_j$ and suppose first that $|\mathrm{Re}\,\xi| \geqslant (r/2)|\mathrm{Im}\,\xi|$. Then we have

$$\begin{aligned}\varphi(\xi) &\leqslant H_K(\xi) + \sigma|\xi| \leqslant H_K(\xi) + \sigma|\mathrm{Re}\,\xi| + \sigma|\mathrm{Im}\,\xi| \\ &\leqslant H_K(\xi) + \varepsilon|\mathrm{Re}\,\xi| \\ &\leqslant H_{K'}(\xi)\,.\end{aligned}$$

Thus there is nothing to prove in this case.

Let now $\xi \in W_j$ be such that $|\mathrm{Re}\,\xi| < (r/2)|\mathrm{Im}\,\xi|$ so that for $\Theta = |\mathrm{Im}\,\xi|^{-1}\xi$ we have $|\mathrm{Im}\,\Theta| = 1$, $|\mathrm{Re}\,\Theta| < r/2$.

We choose a homogeneous polynomial q_j in $\mathbb{C}^n$ with the properties

$$q_j|S(W_j) = 0 \qquad q_j \not\equiv 0 \text{ on } W_j\,.$$

We may also assume that

$$|q_j(\eta)| \leqslant 1 \qquad \text{for } |\eta| \leqslant 3$$

without loss of generality. We make the additional assumption $q_j(\xi) \neq 0$. We consider for $t > 0$ the function

$$\psi_t(\eta) = |\mathrm{Im}\,\xi|^{-1}\varphi(\eta|\mathrm{Im}\,\xi|) + t\log|q(\eta)| \qquad \text{for } \eta \in W_j \cap B(\Theta, r)\,.$$

We have

$$\psi_t(\eta) \leqslant H_K(\eta) + \delta|\eta| \qquad \text{for } \eta \in W_j \cap B(\Theta, r)$$

$$\psi_t(\eta) \leqslant 0 \qquad \text{for } \eta \in W_j \cap i\mathbb{R}^n \cap B(\Theta, r)$$

because on $W_j \cap B(\Theta, r)$ we have $\psi_t(\eta) \leqslant |\mathrm{Im}\,\xi|^{-1} \varphi(\eta|\mathrm{Im}\,\xi|) \leqslant 0$ if $\eta \in W_j \cap \cap\, i\mathbb{R}^n \cap B(\Theta, r)$ and $\eta \notin S(W_j)$, and because $\psi_t(\eta) = -\infty$ if $\eta \in S(W_j) \cap \cap\, B(\Theta, r)$.

Because of condition II(W_j, δ) we must have

$$\psi_t(\Theta) = |\mathrm{Im}\,\xi|^{-1}\varphi(\xi) + t \log\left|q(|\mathrm{Im}\xi|^{-1}\xi)\right| \leqslant |\mathrm{Im}\,\xi|^{-1} H_{K'}(\xi)$$

for every $t > 0$. Since $q(\xi) \neq 0$ we obtain for $t \to 0$ that

$$\varphi(\xi) \leqslant H_{K'}(\xi)$$

for every $\xi \in W_j \cap \{q(\xi) \neq 0\}$.

By the same argument used at the end of the proof of corollary 1 we obtain then that

$$\varphi(\xi) \leqslant H_{K'}(\xi) \qquad \forall \xi \in W_j\,.$$

This proves our contention since j can be taken equal to $1, 2, \ldots, l$.

d) Given the algebraic variety W in $\mathbb{C}^n$ an algebraic subvariety $\Gamma \subset W$ is said to be thin if, for every irreducible component W_j of W, $\Gamma \cap W_j$ is a subvariety of codimension $\geqslant 1$ in W_j.

LEMMA 12. *Given the algebraic cone W in $\mathbb{C}^n$ we can find an algebraic subvariety Γ of W which is thin and contains the singular locus of W, having the following property:*

Given K compact convex in $\mathbb{R}^n$, given $\delta > 0$, $R > 0$ and given $\Theta \in \mathbb{C}^n$ for any weakly plurisubharmonic function φ defined on $W \cap B(\Theta, R)$ and such that

$$\varphi(\xi) \leqslant H_K(\xi) + \delta|\xi| \qquad \forall \xi \in W \cap B(\Theta, R)$$

$$\varphi(\xi) \leqslant 0 \qquad \forall \xi \in W \cap i\mathbb{R}^n \cap B(\Theta, R)$$

we can find a sequence $\{U_n\}_{n \in \mathbb{N}}$ of open neighborhoods of $W \cap B(\Theta, R)$ in $\mathbb{C}^n$ and a corresponding sequence of plurisubharmonic functions $\psi_n \colon U_n \to \mathbb{R} \cup \{-\infty\}$

such that

$$\psi_n(\xi) \leqslant H_K(\xi) + \delta|\xi| \qquad \forall \xi \in U_n$$

$$\psi_n(\xi) \leqslant 0 \qquad \forall \xi \in W \cap i\mathbb{R}^n \cap B(\Theta, R)$$

and such that

$$\varphi(\xi) = \lim_{n\to\infty} \psi_n(\xi) \qquad \forall \xi \in W \cap B(\Theta, R) - \Gamma \cap B(\Theta, R) .$$

PROOF. α) Let $W = W_1 \cup \ldots \cup W_l$ be a decomposition of W into irreducible components. We want first to show that *if the lemma is verified by each irreducible cone W_j ($1 \leqslant j \leqslant l$) then it will be also verified by the cone W itself.*

For each j with $1 \leqslant j \leqslant l$, we consider on W_j the set

$$A_j = W_j \cap \Big(\bigcup_{i\neq j} W_i\Big),$$

and the function φ restricted to $(W_j - A_j) \cap B(\Theta, R)$,

$$\varphi\colon (W_j - A_j) \cap B(\Theta, R) \to \mathbb{R} \cup \{-\infty\} .$$

We set by definition, for $\xi \in W_j \cap B(\Theta, R)$,

$$\varphi_j(\xi) = \max \lim_{\substack{w\to\xi \\ w\in W_j - A_j}} \varphi(w) .$$

Since for every $\xi \in W \cap B(\Theta, R)$ we have $\varphi(\xi) = \max \lim_{\substack{w\to\xi \\ w\in W}} \varphi(w)$ we deduce that, for every $\xi \in W_j \cap B(\Theta, R)$ we have

$$\varphi_j(\xi) \leqslant \varphi(\xi) .$$

In particular the function φ_j satisfies the following conditions

$$\varphi_j(\xi) \leqslant H_K(\xi) + \delta|\xi| \qquad \forall \xi \in W_j \cap B(\Theta, R)$$

$$\varphi_j(\xi) \leqslant 0 \qquad \forall \xi \in W_j \cap i\mathbb{R}^n \cap B(\Theta, R) .$$

Moreover φ_j is weakly plurisubharmonic on $W_j \cap B(\Theta, R)$.

Indeed let $\pi\colon W^* \to W$ denote the normalization of W and let $\pi_j\colon W_j^* \to W_j$ denote the normalization of W_j so that $W^* = W_1^* \dot\cup \ldots \dot\cup W_l^*$.

Let φ^* be the plurisubharmonic function defined by φ on $\pi^{-1}(W \cap \cap B(\Theta, R))$ via the Riemann extension theorem and let $\varphi_j^* = \varphi^* | W_j^* \cap \cap \pi_j^{-1}(W_j \cap B(\Theta, R))$.

Then φ_j^* is plurisubharmonic and, as one verifies, we have

$$\varphi_j(\xi) = \sup\{\varphi_j^*(\xi^*)\,|\,\xi^* \in W_j^*,\ \pi_j(\xi^*) = \xi\}$$

for every $\xi \in W_j \cap B(\Theta, R)$. This proves our contention.

Let us assume now that the lemma is verified by the cone W_j. We can find a proper algebraic subvariety Γ_j of W_j containing the singularities of W_j, a sequence $\{U_n^{(j)}\}_{n\in\mathbb{N}}$ of open neighborhoods in $\mathbb{C}^n$ of $W_i \cap B(\Theta, R)$, and on each $U_n^{(j)}$ a plurisubharmonic function $\psi_n^{(j)}$ such that

$$\begin{aligned} &\psi_n^{(j)}(\xi) \leqslant H_K(\xi) + \delta|\xi| && \forall \xi \in U_n^{(j)} \\ &\psi_n^{(j)}(\xi) \leqslant 0 && \forall \xi \in W_j \cap i\mathbb{R}^n \cap B(\Theta, R) \end{aligned}$$

and such that $\forall \xi \in (W_j - \Gamma_j) \cap B(\Theta, R)$

$$\varphi_j(\xi) = \lim_{n\to\infty} \psi_n^{(j)}(\xi)\,.$$

Let $q_1^{(j)}, \ldots, q_{k_j}^{(j)}$ be a set of (homogeneous) polynomials whose common zeros is the cone $\bigcup_{i\neq j} W_i$. We set

$$\begin{aligned} c_j &= \sup_{B(\Theta,R)} \sum_{h=1}^{k_j} |q_h^{(j)}(\xi)| \\ \phi_j(\xi) &= \log\left(\frac{1}{c_j}\sum_h |q_h^{(j)}(\xi)|\right). \end{aligned}$$

The function ϕ_j is plurisubharmonic (cf. [14] Corollary 1.6.8).

Moreover we have

$$\begin{aligned} &\phi_j(\xi) \leqslant 0 && \forall \xi \in B(\Theta, R) \\ &\phi_j(\xi) = -\infty && \forall \xi \in \bigcup_{i\neq j} W_j\,. \end{aligned}$$

We define $M = \sup_{\xi\in B(\Theta,R)} \{H_K(\xi) + \delta|\xi|\}$ and we set

$$\begin{aligned} {}'G_n^{(j)} &= U_n^{(j)} \cap B(\Theta, R) \\ {}''G_n^{(j)} &= \{\xi \in B(\Theta, R)\,|\,\phi_j(\xi) < -Mn - n^2\}\,. \end{aligned}$$

Since ϕ_j is upper semicontinuous the set ${}''G_n^{(j)}$ is an open set and contains the set $\left(\bigcup_{i\neq j} W_i\right) \cap B(\Theta, R)$ and therefore

$$G_n^{(j)} = {}'G_n^{(j)} \cup {}''G_n^{(j)}$$

is an open neighborhood in $\mathbb{C}^n$ of $W \cap B(\Theta, R)$.

Let

$$'g_n^{(j)}(\xi) = \max\left\{\psi_n^{(j)}(\xi) + \frac{1}{n}\phi_j(\xi), -n\right\} \qquad \text{for } \xi \in 'G_n^{(j)}$$

$$''g_n^{(j)}(\xi) = -n \qquad \text{for } \xi \in ''G_n^{(j)} .$$

On $'G_n^{(j)} \cap ''G_n^{(j)}$ we have

$$\psi_n^{(j)}(\xi) + \frac{1}{n}\phi_j(\xi) \leqslant H_K(\xi) + \delta|\xi| + \frac{1}{n}\phi_j(\xi)$$
$$< H_K(\xi) + \delta|\xi| - M - n \leqslant -n .$$

Therefore on $'G_n^{(j)} \cap ''G_n^{(j)}$ we have $'g_n^{(j)}(\xi) = ''g_n^{(j)}(\xi)$ and we can define on $G_n^{(j)}$

$$g_n^{(j)}(\xi) = \begin{cases} 'g_n^{(j)}(\xi) & \text{for } \xi \in 'G_n^{(j)} \\ ''g_n^{(j)}(\xi) & \text{for } \xi \in ''G_n^{(j)} . \end{cases}$$

This is a plurisubharmonic function on the whole set $G_n^{(j)}$.

There is an integer $n_0 = n_0(K, \delta, \Theta, R)$ such that, for $n \geqslant n_0$ we have

$$(*) \qquad \begin{cases} g_n^{(j)}(\xi) \leqslant H_K(\xi) + \delta|\xi| & \forall \xi \in G_n^{(j)} \\ g_n^{(j)}(\xi) \leqslant 0 & \forall \xi \in W \cap i\mathbb{R}^n \cap B(\Theta, R) . \end{cases}$$

The second inequality is a consequence of the following facts: on $\bigcup_{i \neq j} W_i$ we have $\phi_j = -\infty$ so that on $\left(\bigcup_{i \neq j} W_i\right) \cap B(\Theta, R)$ we must have $g_n^{(j)}(\xi) = -n < 0$; on $W_j \cap i\mathbb{R}^n \cap B(\Theta, R)$ we have $\phi_j(\xi) \leqslant 0$ and $\psi_n^{(j)}(\xi) \leqslant 0$, therefore again $g_n^{(j)}(\xi) \leqslant 0$.

We claim that the first inequality holds for $n \geqslant n_0$ provided

$$-n_0 < \inf_{\xi \in B(\Theta, R)} \{H_K(\xi) + \delta|\xi|\} .$$

Indeed on $''G_n^{(j)}$ we have

$$g_n^{(j)}(\xi) = -n \leqslant -n_0 < H_K(\xi) + \delta|\xi| .$$

On $'G_n^{(j)}$ we have

$$\psi_n^{(j)}(\xi) + \frac{1}{n}\phi_j(\xi) \leqslant \psi_n^{(j)}(\xi) \leqslant H_K(\xi) + \delta|\xi| .$$

Therefore the first inequality holds on $G_n^{(j)}$.

We note now that

$$\lim_{n\to\infty} g_n^{(j)}(\xi) = \begin{cases} \lim_{n\to\infty} \psi_n^{(j)}(\xi) = \varphi_j(\xi) & \text{for } \xi \in (W_j - \Gamma_j) \cap B(\Theta, R) \\ -\infty \quad \text{for } \xi \in \left(\bigcup_{i\neq j} W_i\right) \cap B(\Theta, R) . \end{cases}$$

We define

$$U_n = G_n^{(1)} \cap \ldots \cap G_n^{(l)}$$

$$\psi_n = \sup \{g_n^{(1)}, \ldots, g_n^{(l)}\} \qquad \text{on } U_n .$$

Since each $G_n^{(j)}$ is an open neighborhood in $\mathbb{C}^n$ of $W \cap B(\Theta, R)$ the same is true for U_n. Since each function $g_n^{(j)}$ is plurisubharmonic on $G_n^{(j)}$ we deduce that ψ_n is plurisubharmonic on U_n.

Moreover for $n \geqslant n_0 = n_0(K, \delta, \Theta, R)$ we do have

$$\psi_n(\xi) \leqslant H_K(\xi) + \delta|\xi| \qquad \text{for } \xi \in U_n$$

$$\psi_n(\xi) \leqslant 0 \qquad \text{for } \xi \in W \cap i\mathbb{R}^n \cap B(\Theta, R) .$$

This because of the relations (*) established above.

Set

$$\Gamma = \Gamma_1 \cup \ldots \cup \Gamma_l \cup \left(\bigcup_{i\neq j} (W_i \cap W_j)\right) .$$

This is a thin algebraic subvariety of W which contains the singularities of W.

Let now $\xi \in (W - \Gamma) \cap B(\Theta, R)$, then there exists only one index j such that $\xi \in W_j$. Therefore if $i \neq j$ we have

$$\lim_{n\to\infty} g_n^{(i)}(\xi) = -\infty .$$

Taking into account the fact that $\xi \in W_j - \Gamma_j$ we deduce then

$$\lim_{n\to\infty} \psi_n(\xi) = \lim_{n\to\infty} g_n^{(j)}(\xi) = \lim_{n\to\infty} \psi_n^{(j)}(\xi) = \varphi_j(\xi) = \varphi(\xi) .$$

This proves that the lemma holds on W as we wanted.

β) We assume from now on that W is irreducible; this is not restrictive because of what we proved in point α). Let $d = \dim_{\mathbb{C}} W$.

By $\mathfrak{J}(W)$ we will denote the homogeneous ideal of homogeneous polynomials defined in $\mathbb{C}^n$, and vanishing on W. We may assume to be in the conditions of the preparation lemma 8 for W. In particular we may assume that

(i) $\mathfrak{J}(W) \cap \mathbb{C}[\xi_1, \ldots, \xi_d] = 0$

(ii) for every j with $1 \leqslant j \leqslant n-d$ the ideal

$$\mathfrak{J}(W) \cap \mathbf{C}[\xi_1, \ldots, \xi_d, \xi_{d+j}] = Q_j(\xi_1, \ldots, \xi_d, \xi_{d+j})\mathbf{C}[\xi_1, \ldots, \xi_d, \xi_{d+j}]$$

is a prime principal ideal generated by the irreducible homogeneous polynomial Q_j which we may also suppose monic with respect to ξ_{d+j}. We denote by $\Delta_j(\xi_1, \ldots, \xi_d)$ the discriminant with respect to ξ_{d+j} of Q_j and we set

$$\Delta(\xi_1, \ldots, \xi_d) = \Delta_1 \ldots \Delta_{n-d} .$$

For $\xi_1, \ldots, \xi_d$ fixed in $\mathbf{C}^d$, we denote by

$$\xi_{d+j}^{(1)}, \ldots, \xi_{d+j}^{(m_j)}$$

the roots (counted with their multiplicity) of the equation $Q_j(\xi_1, \ldots, \xi_d, t) = 0$; here m_j denotes the degree of Q_j: we have thus

$$\Delta_j(\xi_1, \ldots, \xi_d) = \prod_{r<s} (\xi_{d+j}^{(r)} - \xi_{d+j}^{(s)})^2 .$$

Also from the preparation lemma (condition (ξ)) we derive that there is a constant $c > 0$ such that

$$|\xi_{d+j}^{(s)}| \leqslant c\Big(\sum_1^d |\xi_i|^2\Big)^{\frac{1}{2}} .$$

Thus for r, s fixed between 1 and m_j, with $r < s$, we obtain

$$|\Delta_j(\xi_1, \ldots, \xi_d)| \leqslant |\xi_{d+j}^{(r)} - \xi_{d+j}^{(s)}|^2 (2c)^{2(m_j-1)} \Big(\sum_1^d |\xi_i|^2\Big)^{m_j-1}$$
$$\leqslant (2c)^{2m_j} \Big(\sum_1^d |\xi_i|^2\Big)^{m_j} .$$

From these estimates we derive

$$|\Delta(\xi_1, \ldots, \xi_d)| \leqslant |\xi_{d+j}^{(r)} - \xi_{d+j}^{(s)}|^2 a(\xi)$$

with

$$a(\xi) = (2c)^{2\sum_1^{n-d} m_k - 2} \Big(\sum_1^d |\xi_i|^2\Big)^{\sum_1^{n-d} m_k - 1} .$$

Given $\Theta \in \mathbb{C}^n$, $R > 0$, set $|\Theta| + R = \varrho$ so that for $\xi \in B(\Theta, R)$ we certainly have $\sum_1^d |\xi_i|^2 \leqslant \varrho^2$.

Set

$$c_0 = \sup_{\sum_1^d |\xi_i|^2 \leqslant \varrho^2} a(\xi) .$$

We conclude that *for* $\sum_1^d |\xi_i|^2 \leqslant \varrho^2$ *for any two roots* $\xi_{d+j}^{(r)}$ *and* $\xi_{d+j}^{(s)}$ *with* $r < s$ *of* $Q_j(\xi_1, \ldots, \xi_d, t) = 0$ *we have*

$$(*) \qquad |\xi_{d+j}^{(r)} - \xi_{d+j}^{(s)}|^2 \geqslant \frac{|\Delta(\xi_1, \ldots, \xi_d)|}{c_0} .$$

We also set

$$c_1 = 1 + \sup_{\sum_1^d |\xi_i|^2 \leqslant \varrho^2} |\Delta(\xi_1, \ldots, \xi_d)| .$$

We now consider the following set

$$U = \bigcup_{\substack{w \in W \cap B(\Theta, R) \\ \Delta(w) \neq 0}} \left\{\xi \in B(\Theta, R) | \xi_i = w_i \text{ for } 1 \leqslant i \leqslant d \text{ and } |\xi - w| < \frac{1}{2} \sqrt{\frac{|\Delta(w)|}{c_0}}\right\} .$$

Then U is an open neighborhood in $\mathbb{C}^n$ of the set

$$Z = \{w \in W \cap B(\Theta, R) | \Delta(w) \neq 0\} .$$

Now we remark that because of the inequality $(*)$ given above, for any $\xi \in U$ there exists a unique point $w \in Z$ such that $\xi_i = w_i$ for $1 \leqslant i \leqslant d$. Moreover the other coordinates $w_{d+1}, \ldots, w_n$ of w are holomorphic functions of the first d.

This shows that we have a *holomorphic retraction*

$$\pi \colon U \to Z$$

$(\pi(w) = w \;\; \forall w \in Z)$.

Let now φ be a weakly plurisubharmonic function defined on $W \cap \cap B(\Theta, R)$.

If we note that Z consists only of non singular points of W, it follows that $\varphi|Z$ is plurisubharmonic. Therefore also the function

$$\pi^* \varphi = \varphi \circ \pi$$

defined on U is plurisubharmonic in U and extends $\varphi|Z$ to the open set U:

$$\pi^* \varphi(w) = \varphi(w) \qquad \text{for } w \in Z .$$

Let us suppose that φ satisfies the conditions

$$\begin{aligned} &\varphi(\xi) \leqslant H_K(\xi) + \delta|\xi| \qquad && \forall \xi \in W \cap B(\Theta, R) \\ &\varphi(\xi) \leqslant 0 && \forall \xi \in W \cap i\mathbb{R}^n \cap B(\Theta, R) . \end{aligned}$$

We set

$$M = \sup_{B(\Theta,R)} \big|H_K(\xi) + \delta|\xi|\big| + 1 .$$

We consider for $\varepsilon > 0$ the plurisubharmonic function defined on U by

$$\psi_\varepsilon^{(1)}(\xi) = \sup\left\{\pi^*\varphi(\xi) + \varepsilon \log\frac{|\Delta(\xi)|}{c_1}, -\frac{1}{\varepsilon}\right\} .$$

We note that, because of the choice of the constant c_1 we have that $\log|\Delta(\xi)|/c_1 < 0$ for $\xi \in B(\Theta, R)$. Consequently the set

$$U_\varepsilon^{(1)} = \left\{\xi \in U | \pi^*\varphi(\xi) + \frac{\varepsilon}{2}\log\frac{|\Delta(\xi)|}{c_1} < H_K(\xi) + \delta|\xi|\right\}$$

is an open neighborhood of Z in U and thus in $\mathbb{C}^n$.

We also define

$$U_\varepsilon^{(2)} = \left\{\xi \in B(\Theta, R) | \log\frac{|\Delta(\xi)|}{c_1} < -\frac{2M}{\varepsilon} - \frac{2}{\varepsilon^2}\right\}$$

and this is an open neighborhood of the set $A = B(\Theta, R) \cap W \cap \{\Delta(\xi) = 0\}$. Note that $Z \cup A = W \cap B(\Theta, R)$ so that $U_\varepsilon = U_\varepsilon^{(1)} \cup U_\varepsilon^{(2)}$ is an open neighborhood in $\mathbb{C}^n$ of $W \cap B(\Theta, R)$.

On $U_\varepsilon^{(1)} \cap U_\varepsilon^{(2)}$ we do have

$$\begin{aligned} \pi^*\varphi(\xi) + \varepsilon\log\frac{|\Delta(\xi)|}{c_1} &< H_K(\xi) + \delta|\xi| + \frac{\varepsilon}{2}\log\frac{|\Delta(\xi)|}{c_1} \\ &< M - \frac{\varepsilon}{2}\left(\frac{2M}{\varepsilon} + \frac{2}{\varepsilon^2}\right) \\ &= -\frac{1}{\varepsilon} . \end{aligned}$$

We define therefore the following function on U_ε

$$\psi_\varepsilon(\xi) = \begin{cases} \psi_\varepsilon^{(1)}(\xi) & \text{for } \xi \in U_\varepsilon^{(1)} \\ -\dfrac{1}{\varepsilon} & \text{for } \xi \in U_\varepsilon^{(2)} . \end{cases}$$

This will be plurisubharmonic because of the remark made just above. We choose $\varepsilon_0 > 0$ such that

$$-\frac{1}{\varepsilon_0} < \inf_{B(\Theta,R)} (H_K(\xi) + \delta|\xi|)$$

and we consider $\varepsilon > 0$ with $0 < \varepsilon \leqslant \varepsilon_0$. Then we have in the whole of U_ε

$$\psi_\varepsilon(\xi) \leqslant H_K(\xi) + \delta|\xi| \qquad \forall \xi \in U_\varepsilon .$$

Indeed the estimate is satisfied in $U_\varepsilon^{(1)}$ by the definition of the set $U_\varepsilon^{(1)}$ and because $-1/\varepsilon < -1/\varepsilon_0$. For this last reason it is satisfied also in $U_\varepsilon^{(2)}$.

Now if $\xi \in W \cap i\mathbb{R}^n \cap B(\Theta, R)$ we have $\psi_\varepsilon^{(1)}(\xi) \leqslant 0$ on $U_\varepsilon^{(1)}$ because, for that ξ, $\pi^*\varphi(\xi) = \varphi(\xi) \leqslant 0$ and by the choice of c_1 since then $\log |\Delta(\xi)|/c_1 < 0$. On $U_\varepsilon^{(2)}$ the inequality is obvious. Therefore

$$\psi_\varepsilon(\xi) \leqslant 0 \qquad \forall \xi \in W \cap i\mathbb{R}^n \cap B(\Theta, R) .$$

Let now $\xi \in W \cap B(\Theta, R)$ with $\Delta(\xi) \neq 0$ (i.e. $\xi \in Z$). Then

$$\lim_{\varepsilon \to 0} \psi_\varepsilon(\xi) = \lim_{\varepsilon \to 0} \varphi(\xi) + \varepsilon \log \frac{|\Delta(\xi)|}{c_1} = \varphi(\xi)$$

provided $\varphi(\xi) > -\infty$. If $\varphi(\xi) = -\infty$ then $\psi_\varepsilon(\xi) = -1/\varepsilon$ and again

$$\lim_{\varepsilon \to 0} \psi_\varepsilon(\xi) = \lim_{\varepsilon \to 0} -\frac{1}{\varepsilon} = -\infty = \varphi(\xi) .$$

Thus in any case we have for $\xi \in W \cap B(\Theta, R)$ and $\Delta(\xi) \neq 0$

$$\lim_{\varepsilon \to 0} \psi_\varepsilon(\xi) = \varphi(\xi) .$$

We take for Γ the set $\{\xi \in W | \Delta(\xi) = 0\}$ which is a thin algebraic subvariety of W. The lemma is thus satisfied taking a sequence of functions ψ_ε as approximating functions to φ outside Γ.

REMARK. The proof of point β) given above establishes in general that for an irreducible cone W we have:

for every $\xi \in W \cap B(\Theta, R)$ there exists the limit $\lim_{\varepsilon\to 0} \psi_\varepsilon(\xi)$ and that

$$\lim_{\varepsilon\to 0} \psi_\varepsilon(\xi) \leqslant \varphi(\xi) .$$

Indeed when $\Delta(\xi) \neq 0$ this is what we proved, the equality sign being valid in this case. If $\Delta(\xi) = 0$ then we have $\lim_{\varepsilon\to 0} \psi_\varepsilon(\xi) = -\infty$ and the above inequality is certainly valid.

Set $\psi(\xi) = \lim_{\varepsilon\to 0} \psi_\varepsilon(\xi)$ so that we have

$$\psi(\xi) \leqslant \varphi(\xi) \qquad \forall \xi \in W \cap B(\Theta, R)$$

$$\psi(\xi) = \varphi(\xi) \qquad \forall \xi \in (W - \Gamma) \cap B(\Theta, R) .$$

It follows that φ is the smallest upper semicontinuous majorant of ψ. Indeed φ being upper semicontinuous it is an upper semicontinuous majorant of ψ. Moreover for $\xi \in \Gamma \cap B(\Theta, R)$ we have

$$\max \lim_{\substack{w\to\xi \\ w\in W\cap B(\Theta,R)}} \psi(w) \geqslant \max \lim_{\substack{w\to\xi \\ w\in(W-\Gamma)\cap B(\Theta,R)}} \psi(w) = \max \lim_{\substack{w\to\xi \\ w\in(W-\Gamma)\cap B(\Theta,R)}} \varphi(w) = \varphi(\xi) .$$

The last equality follows from Riemann extension theorem applied to the normalization W^* of W. We deduce then that $\forall \xi \in W \cap B(\Theta, R)$

$$\begin{aligned} \varphi(\xi) &= \max \lim_{w\to\xi} \psi(w) \\ &= \max \lim_{w\to\xi} \max \lim_{\varepsilon\to 0} \psi_\varepsilon(w) . \end{aligned}$$

This proves in particular our contention.

As a corollary we obtain the following

PROPOSITION 17. *Let $K \subset K'$ be compact convex sets in $\mathbb{R}^n$ and let $K \subset \mathring{K}'$. Let r with $0 < r < 1$ be given.*

Then the Phragmén-Lindelöf principle (I) *on W is equivalent to the following statement*

(III) « *There is a $\delta > 0$ such that, for any choice of $\Theta \in W$ with $|\mathrm{Im}\, \Theta| = 1$, and for any choice of a plurisubharmonic function $\varphi\colon U \to \to \mathbb{R} \cup \{-\infty\}$ defined in an open neighborhood U of the set $W \cap B(\Theta, r)$*

and such that

$$\varphi(\xi) \leqslant H_K(\xi) + \delta|\xi| \qquad \forall \xi \in U$$

$$\varphi(\xi) \leqslant 0 \qquad \forall \xi \in W \cap i\mathbb{R}^n \cap B(\Theta, r)$$

one has also that

$$\varphi(\Theta) \leqslant H_{K'}(\Theta) \text{ ».}$$

PROOF. We will show that statements (II) and (III) are equivalent.

α) (II) $\Rightarrow$ (III). We distinguish two cases:

Case 1. $|\operatorname{Re}\Theta| < r/2$. Choose $\delta = \sigma$ and let φ be as in statement (III). Then $\varphi|W \cap B(\Theta, r)$ is weakly plurisubharmonic and we are in the assumptions of statement (II). Therefore $\varphi(\Theta) \leqslant H_{K'}(\Theta)$ as we wanted.

Case 2. $|\operatorname{Re}\Theta| \geqslant r/2$. Then

$$H_K(\Theta) + \delta|\Theta| \leqslant H_K(\Theta) + \delta\left(1 + \frac{2}{r}\right)|\operatorname{Re}\Theta| .$$

Since $K \subset \mathring{K}'$ for $\varepsilon > 0$ sufficiently small

$$H_K(\xi) + \varepsilon|\operatorname{Re}\xi| \leqslant H_{K'}(\xi) \qquad \forall \xi \in \mathbb{C}^n .$$

It is enough to choose δ so small that $\delta(1 + 2/r) < \varepsilon$ to be able to deduce the last inequality of condition (III) from the first.

β) (III) $\Rightarrow$ (II). We let r be chosen with $0 < r < 1$ and we select r_1 with $0 < r_1 < r$. We assume that condition (III) holds when r is replaced by r_1. We will show that (II) holds with $\sigma = \delta$.

Let φ be a weakly plurisubharmonic function defined on $W \cap B(\Theta, r)$ and such that

$$\varphi(\xi) \leqslant H_K(\xi) + \delta|\xi| \qquad \forall \xi \in W \cap B(\Theta, r)$$

$$\varphi(\xi) \leqslant 0 \qquad \forall \xi \in W \cap i\mathbb{R}^n \cap B(\Theta, r) .$$

By the previous lemma we can find a sequence of open neighborhoods U_n of $W \cap B(\Theta, r)$ in $\mathbb{C}^n$, a sequence of plurisubharmonic functions ψ_n: $U_n \to \mathbb{R} \cup \{-\infty\}$ and a thin algebraic subvariety $\Gamma \subset W$ (containing the singularities of W) such that

$$\begin{aligned} &\psi_n(\xi) \leqslant H_K(\xi) + \delta|\xi| && \forall \xi \in U_n \\ &\psi_n(\xi) \leqslant 0 && \forall \xi \in W \cap i\mathbb{R}^n \cap B(\Theta, r) \\ &\lim_{n\to\infty} \psi_n(\xi) = \varphi(\xi) && \forall \xi \in (W - \Gamma) \cap B(\Theta, r)\,. \end{aligned}$$

Choose ϱ with $0 < \varrho < (r - r_1)/(1 + r_1)$ and let $w \in W \cap B(\Theta, \varrho)$ be fixed. Since $\varrho < 1$, we have that $\operatorname{Im} w \neq 0$ because

$$1 - |\operatorname{Im} w| \leqslant |\operatorname{Im} \Theta - \operatorname{Im} w| \leqslant |\Theta - w| < \varrho\,.$$

On the other hand we have

$$\begin{aligned} |\operatorname{Im} w| &\leqslant |\operatorname{Im} \Theta| + |\operatorname{Im} w - \operatorname{Im} \Theta| \\ &\leqslant 1 + |w - \Theta| \\ &\leqslant 1 + \varrho\,. \end{aligned}$$

We claim that for $\xi \in B(w/|\operatorname{Im} w|, r_1)$ we have $\big||\operatorname{Im} w|\xi - \Theta\big| < r$. In fact

$$\begin{aligned} \big||\operatorname{Im} w|\xi - \Theta\big| &\leqslant |\Theta - w| + |\operatorname{Im} w| \left|\xi - \frac{w}{|\operatorname{Im} w|}\right| \\ &\leqslant \varrho + (1 + \varrho) r_1 \\ &\leqslant r \end{aligned}$$

by the choice of ϱ.

It follows that the set

$$U_n(w) = \{\xi \in \mathbb{C}^n \big|\ |\operatorname{Im} w|\xi \in U_n\}$$

is open and contains $W \cap B(w/|\operatorname{Im} w|, r_1)$ and therefore it is an open neighborhood of this last set.

The function of η

$$\psi_n(\eta) = |\operatorname{Im} w|^{-1} \psi_n(\eta |\operatorname{Im} w|)$$

is defined and plurisubharmonic on $U_n(w)$ for every n.

Moreover we have

$$\begin{aligned} &\psi_n(\eta) \leqslant H_K(\eta) + \delta|\eta| \quad \text{for } \eta \in U_n(w) \\ &\psi_n(\eta) \leqslant 0 \quad \text{for } \eta \in W \cap i\mathbb{R}^n \cap B\left(\frac{w}{|\operatorname{Im} w|}, r_1\right). \end{aligned}$$

Because of condition (III) we deduce that for $\eta = w/|\mathrm{Im}\, w|$ we have

$$|\mathrm{Im}\, w|^{-1}\psi_n(w) \leqslant H_{K'}\left(\frac{w}{|\mathrm{Im}\, w|}\right) = |\mathrm{Im}\, w|^{-1} H_{K'}(w)$$

and therefore

$$\psi_n(w) \leqslant H_{K'}(w)$$

for every n and every choice of $w \in W \cap B(\Theta, \varrho)$.

Now for $w \in (W - \Gamma) \cap B(\Theta, \varrho)$ we can take the limit for $n \to \infty$ and we obtain

$$\varphi(w) \leqslant H_{K'}(w) \qquad \forall w \in (W - \Gamma) \cap B(\Theta, \varrho) .$$

Now as we have remarked several times we have, since φ is weakly plurisubharmonic and since Γ is thin (1)

$$\varphi(\Theta) = \max \lim_{\substack{\xi \to \Theta \\ \xi \in (W - \Gamma) \cap B(\Theta, r)}} \varphi(\xi) .$$

Therefore from the above inequality we derive

$$\varphi(\Theta) \leqslant H_{K'}(\Theta) .$$

This shows that condition (II) holds.

LEMMA 13. *Let K be compact convex, let δ, δ_1, R, R_1 be real numbers with $0 < \delta < \delta_1$ and $0 < R_1 < R$, let $\Theta \in \mathbb{C}^n$.*

Assume that on an open neighborhood U of $W \cap B(\Theta, R)$ we have defined a plurisubharmonic function $\varphi\colon U \to \mathbb{R} \cup \{-\infty\}$ such that

$$\varphi(\xi) \leqslant H_K(\xi) + \delta|\xi| \qquad \forall \xi \in U$$

$$\varphi(\xi) \leqslant 0 \qquad \forall \xi \in W \cap i\mathbb{R}^n \cap B(\Theta, R) .$$

(1) Let Ω be open in $\mathbb{C}^n$, let Z be an analytic subset of Ω and let $\varphi\colon Z \to \mathbb{R} \cup \{-\infty\}$ be a weakly plurisubharmonic function defined on Z. Let $A \subset Z$ be a thin subset of Z (i.e. of codimension $\geqslant 1$ on every irreducible germ of Z). For any point $a \in A$ we have

$$\varphi(a) = \max \lim_{\substack{z \to a \\ z \in Z - A}} \varphi(z) .$$

This follows from Riemann extension theorem on the normalization Z^* of Z (cf. [11], Satz. 1).

Then we can find a plurisubharmonic function defined on $B(\Theta, R_1)$, $\psi\colon B(\Theta, R_1) \to \mathbb{R} \cup \{-\infty\}$, *satisfying the conditions*

$$\psi(\xi) \leqslant H_K(\xi) + \delta_1|\xi| \qquad \forall \xi \in B(\Theta, R_1)$$
$$\psi(\xi) = \varphi(\xi) \qquad \forall \xi \in W \cap B(\Theta, R_1)\,.$$

PROOF. Let $q_1(\xi), \ldots, q_s(\xi)$ be a set of homogeneous polynomials whose set of common zeros is the cone W.

Since U is an open neighborhood of $\overline{W \cap B(\Theta, R)}$ we derive that the set $\overline{W \cap B(\Theta, R_1)}$ is a compact subset of U and thus disjoint from the compact set $\overline{B(\Theta, R_1)} \cap C(U)$. We can therefore find $\varepsilon > 0$ so small that

$$\left\{\xi \in B(\Theta, R_1) \,|\, \sum |q_j(\xi)| < 2\varepsilon\right\} \subset U\,.$$

Since the q_j's are homogeneous polynomials for every $\varepsilon > 0$ we can find a constant $c_\varepsilon > 0$ such that

$$\frac{1}{\varepsilon}\sum |q_j(\xi)| \leqslant \exp\left(c_\varepsilon|\xi|\right) \qquad \forall \xi \in \mathbb{C}^n\,.$$

We now define

$$\psi^{(1)}(\xi) = \sup\left\{\varphi(\xi),\, H_K(\xi) + \delta|\xi| + \frac{\delta_1 - \delta}{c_\varepsilon}\log\frac{\sum |q_j(\xi)|}{\varepsilon}\right\}$$

for $\xi \in B(\Theta, R_1)$ and $\sum |q_j(\xi)| < 2\varepsilon$;

$$\psi^{(2)}(\xi) = H_K(\xi) + \delta|\xi| + \frac{\delta_1 - \delta}{c_\varepsilon}\log\frac{\sum |q_j(\xi)|}{\varepsilon}$$

for $\xi \in B(\Theta, R_1)$ and $\sum |q_j(\xi)| > \varepsilon$.

When $\xi \in B(\Theta, R_1)$ and $\varepsilon < \sum |q_j(\xi)| < 2\varepsilon$ we do have that $\log \sum |q_j(\xi)|/\varepsilon > 0$ and therefore (since $\xi \in U$)

$$\varphi(\xi) \leqslant H_K(\xi) + \delta|\xi| \leqslant H_K(\xi) + \delta|\xi| + \frac{\delta_1 - \delta}{c_\varepsilon}\log\frac{\sum |q_j(\xi)|}{\varepsilon}\,.$$

Therefore in this set we have $\psi^{(1)}(\xi) = \psi^{(2)}(\xi)$ and the functions $\psi^{(i)}$ $i = 1, 2$, define a unique plurisubharmonic function ψ on $B(\Theta, R_1)$.

Since on W we have $\log \sum |q_j(\xi)|/\varepsilon = -\infty$ we deduce that

$$\psi(\xi) = \varphi(\xi) \;\; \forall \xi \in W \cap B(\Theta, R_1)\,.$$

Now we have $\forall \xi \in \mathbb{C}^n$

$$H_K(\xi) + \delta|\xi| + \frac{\delta_1 - \delta}{c_\varepsilon}\log\frac{\sum |q_j(\xi)|}{\varepsilon} \leqslant H_K(\xi) + \delta_1|\xi|\,.$$

This implies that we have

$$\psi(\xi) \leqslant H_K(\xi) + \delta_1|\xi| \qquad \forall \xi \in B(\Theta, R_1) .$$

As a corollary we obtain the following

PROPOSITION 18. *Let $K \subset K'$ compact convex sets in $\mathbb{R}^n$ with $K \subset \mathring{K}'$. Then the Phragmén-Lindelöf principle* (I) *on W is equivalent to the following statement*

(IV) « *Given r with $0 < r < 1$ there is a $\delta > 0$ such that for any choice of $\Theta \in W$ with $|\mathrm{Im}\, \Theta| = 1$ and for any choice of a plurisubharmonic function φ defined on $B(\Theta, r)$ and satisfying the conditions*

$$\varphi(\xi) \leqslant H_K(\xi) + \delta|\xi| \qquad \forall \xi \in B(\Theta, r)$$

$$\varphi(\xi) \leqslant 0 \qquad \forall \xi \in W \cap i\mathbb{R}^n \cap B(\Theta, r)$$

we have also

$$\varphi(\Theta) \leqslant H_{K'}(\Theta) \text{ ».}$$

PROOF. We will prove that (IV) is equivalent to (III).

The implication (III) $\Rightarrow$ (IV) is obvious.

We prove that (IV) $\Rightarrow$ (III). Choose r_1 with $0 < r_1 < r$ and let $\delta > 0$ be such that (IV) holds for r_1 instead of r and for δ. Let us choose δ_1 with $0 < \delta_1 < \delta$ and let φ be a plurisubharmonic function defined in a neighborhood U of $W \cap B(\Theta, r)$ and satisfying the assumption of condition (III) with δ_1 replacing δ, so that

$$\varphi(\xi) \leqslant H_K(\xi) + \delta_1|\xi| \qquad \forall \xi \in U$$

$$\varphi(\xi) \leqslant 0 \qquad \forall \xi \in W \cap i\mathbb{R}^n \cap B(\Theta, r) .$$

By the previous lemma we can construct ψ plurisubharmonic on $B(\Theta, r_1)$ such that

$$\varphi(\xi) = \psi(\xi) \qquad \forall \xi \in W \cap B(\Theta, r_1)$$

and such that

$$\psi(\xi) \leqslant H_K(\xi) + \delta|\xi| \qquad \forall \xi \in B(\Theta, r_1)$$

$$\psi(\xi) \leqslant 0 \qquad \forall \xi \in W \cap i\mathbb{R}^n \cap B(\Theta, r_1) .$$

Because of condition (IV) (for r_1 and δ) we derive

$$\varphi(\Theta) = \psi(\Theta) \leqslant H_{K'}(\Theta) .$$

Therefore condition (III) holds (for r and δ_1).

LEMMA 14. *Let K be convex compact in $\mathbb{R}^n$, let r with $0 < r < 1$ and $\delta > 0$ be given.*

We can find $\delta_1 > 0$ such that for any choice of $\Theta \in W$ with $|\mathrm{Im}\,\Theta| = 1$, $|\mathrm{Re}\,\Theta| < r/2$ and for any choice of a plurisubharmonic function φ defined on $B(\Theta, r)$ and satisfying

$$\varphi(\xi) \leqslant H_K(\xi) + \delta_1|\xi| \qquad \forall \xi \in B(\Theta, r)$$

$$\varphi(\xi) \leqslant 0 \qquad \forall \xi \in W \cap i\mathbb{R}^n \cap B(\Theta, r)$$

there exists a plurisubharmonic function ψ defined on $\mathbb{C}^n$ with the following properties:

$$\psi(\xi) \leqslant H_K(\xi) + \delta|\xi| \qquad \forall \xi \in \mathbb{C}^n$$

$$\psi(\xi) \leqslant 0 \qquad \forall \xi \in W \cap i\mathbb{R}^n$$

$$\psi(\Theta) \geqslant \varphi(\Theta)$$

$$\psi(\xi) = H_K(\xi) + \delta|\mathrm{Re}\,\xi| \qquad \forall \xi \in \mathbb{C}^n - B(\Theta, r) .$$

PROOF. The proof is similar to the proof of lemma 11.

From the proof of that lemma we borrow the definition of the plurisubharmonic ϕ on $\mathbb{C}^n$ (point α)). We recall that ϕ has the properties

i) $\phi(\xi) < |\mathrm{Re}\,\xi| \ \forall \xi \in \mathbb{C}^n - \{0\}$

ii) given $r_2 > r_1 > 0$ there exists a $\mu > 0$ such that

$$\phi(\xi) < |\mathrm{Re}\,\xi| - \mu \qquad \text{for } r_1 \leqslant |\xi| \leqslant r_2$$

iii) $\phi(\xi) > 0 \ \forall \xi \in \mathbb{R}^n$.

We note that we have (see the proof of lemma 11 point β)) that for ξ near and on the boundary of $B(\Theta, r)$ we have

$$\frac{r}{2} < |\xi - i\,\mathrm{Im}\,\Theta| < \frac{3r}{2} .$$

Taking $r_1 = r/2$ and $r_2 = 3r/2$ we find $\mu > 0$ so that

$$\phi(\xi) < |\mathrm{Re}\,\xi| - \mu \qquad \text{for } \frac{r}{2} \leqslant |\xi| \leqslant \frac{3r}{2}.$$

We then choose δ_1 with $0 < \delta_1 < \min\{\delta/2, \delta\mu/6\}$.

We define the plurisubharmonic function

$$\psi_1(\xi) = \sup\left\{\varphi(\xi) + \frac{\delta}{2}\phi(\xi - i\,\mathrm{Im}\,\Theta), H_K(\xi) + \delta|\mathrm{Re}\,\xi|\right\}$$

for $\xi \in B(\Theta, r)$.

We have for $\xi \in B(\Theta, r)$

$$(*) \qquad \varphi(\xi) + \frac{\delta}{2}\phi(\xi - \mathrm{Im}\,\Theta) \leqslant H_K(\xi) + \delta_1|\xi| + \frac{\delta}{2}|\mathrm{Re}\,\xi| \leqslant H_K(\xi) + \delta|\xi|$$

since $\delta_1 < \delta/2$.

Also for $\xi \in B(\Theta, r)$ we have $|\xi| < 1 + r/2 + r < 3$. If we restrict ξ to a neighborhood of $\partial B(\Theta, r)$ we deduce that then

$$\begin{aligned}\varphi(\xi) + \frac{\delta}{2}\phi(\xi - i\,\mathrm{Im}\,\Theta) &\leqslant H_K(\xi) + \frac{\delta}{2}|\mathrm{Re}\,\xi| + 3\delta_1 - \frac{\delta\mu}{2}\\ &< H_K(\xi) + \frac{\delta}{2}|\mathrm{Re}\,\xi|\\ &< H_K(\xi) + \delta\,|\mathrm{Re}\,\xi|.\end{aligned}$$

We can then define a global plurisubharmonic function ψ on $\mathbf{C}^n$ by setting

$$\psi(\xi) = \begin{cases} \psi_1(\xi) & \text{for } \xi \in B(\Theta, r)\\ H_K(\xi) + \delta|\mathrm{Re}\,\xi| & \text{for } \xi \in \mathbf{C}^n - B(\Theta, r).\end{cases}$$

We have for any $\xi \in \mathbf{C}^n$

$$\psi(\xi) \leqslant H_K(\xi) + \delta|\xi|$$

because of $(*)$.

Also

$$\psi(\xi) \leqslant 0 \qquad \text{if } \xi \in W \cap i\mathbb{R}^n$$

because for $\xi \in W \cap i\mathbb{R}^n \cap B(\Theta, r)$ we have

$$\varphi(\xi) + \frac{\delta}{2}\phi(\xi - i\,\mathrm{Im}\,\Theta) \leqslant \varphi(\xi) \leqslant 0$$

(by property iii) of φ), while for $\xi \in i\mathbb{R}^n$ $H_K(\xi) + \delta|\mathrm{Re}\,\xi| = 0$. Finally for $\xi = \Theta$ we derive

$$\psi(\Theta) \geq \varphi(\Theta) + \frac{\delta}{2}\phi(\mathrm{Re}\,\Theta) \geq \varphi(\Theta)$$

(again by property iii) of ϕ). This achieves the proof.

As a consequence we derive the following

PROPOSITION 19. *Let $K \subset K'$ compact convex sets in $\mathbb{R}^n$ with $K \subset \mathring{K}'$.*

The Phragmén-Lindelöf principle (I) *on W is equivalent to the following statement*

(V) « *There is a $\delta > 0$ such that for any choice of a plurisubharmonic function φ defined on the whole of $\mathbb{C}^n$ and satisfying*

$$\varphi(\xi) \leq H_K(\xi) + \delta|\xi| \qquad \forall \xi \in \mathbb{C}^n$$

$$\varphi(\xi) \leq 0 \qquad \forall \xi \in W \cap i\mathbb{R}^n$$

we have also

$$\varphi(\xi) \leq H_{K'}(\xi) \qquad \forall \xi \in W \text{ ».}$$

PROOF. We will prove that (I) $\Rightarrow$ (V) $\Rightarrow$ (IV). By the equivalence of (I) and (IV) the proposition will be proved.

The implication (I) $\Rightarrow$ (V) is obvious since the restriction to W of a plurisubharmonic function in $\mathbb{C}^n$ is certainly weakly plurisubharmonic.

Assume now that (V) holds as stated. Choose r with $0 < r < 1$ and choose δ_1 with $0 < \delta_1 < \delta$ such that lemma 14 holds.

Let $\Theta \in W$ with $|\mathrm{Im}\,\Theta| = 1$ and let φ be plurisubharmonic in $B(\Theta, r)$ with

$$\varphi(\xi) \leq H_K(\xi) + \delta_1|\xi| \qquad \forall \xi \in B(\Theta, r)$$

$$\varphi(\xi) \leq 0 \qquad \forall \xi \in W \cap i\mathbb{R}^n \cap B(\Theta, r)\,.$$

We distinguish two cases

Case 1. $|\mathrm{Re}\,\Theta| \geq r/2$ then

$$H_K(\Theta) + \delta_1|\Theta| \leq H_K(\Theta) + \delta_1\left(1 + \frac{2}{r}\right)|\mathrm{Re}\,\Theta|$$

and since $K \subset \mathring{K}'$ there is an $\varepsilon > 0$ such that

$$H_K(\xi) + \varepsilon|\mathrm{Re}\,\xi| \leq H_{K'}(\xi) \qquad \forall \xi \in \mathbb{C}^n\,.$$

Therefore if $\delta_1(1 + 2/r) < \varepsilon$ we do have from the first inequality

$$\varphi(\Theta) \leqslant H_{K'}(\Theta)$$

as we wanted (provided δ_1 has been chosen sufficiently small).

Case 2. $|\mathrm{Re}\,\Theta| < r/2$. By the previous lemma 14 we can find a plurisubharmonic function ψ on $\mathbb{C}^n$ with

$$\begin{aligned} &\psi(\xi) \leqslant H_K(\xi) + \delta|\xi| && \forall \xi \in \mathbb{C}^n \\ &\psi(\xi) \leqslant 0 && \forall \xi \in W \cap i\mathbb{R}^n \\ &\psi(\Theta) \geqslant \varphi(\Theta) \,. \end{aligned}$$

Since (V) holds we do have

$$\psi(\xi) \leqslant H_{K'}(\xi) \qquad \forall \xi \in W;$$

in particular

$$\varphi(\Theta) \leqslant \psi(\Theta) \leqslant H_{K'}(\Theta)$$

as we wanted.

g) Lemma 15. *Let K be a compact convex set in $\mathbb{R}^n$, let δ, r, a be real numbers with $\delta > 0$, $0 < r < 1$, $a > 0$.*

We can find $\delta_1 > 0$ such that, for any choice of $\Theta \in W$ with $|\mathrm{Im}\,\Theta| = 1$, $|\mathrm{Re}\,\Theta| < r/2$, for any choice of a plurisubharmonic function φ defined in $B(\Theta, r)$ and such that

$$\begin{aligned} &\varphi(\xi) \leqslant H_K(\xi) + \delta_1|\xi| && \forall \xi \in B(\Theta, r) \\ &\varphi(\xi) \leqslant 0 && \forall \xi \in W \cap i\mathbb{R}^n \cap B(\Theta, r) \,, \end{aligned}$$

for any choice of $\varepsilon > 0$ we can find a plurisubharmonic function ψ on $\mathbb{C}^n$ with the properties

$$\begin{aligned} &\psi(\xi) \leqslant H_K(\xi) + \delta|\xi| && \forall \xi \in \mathbb{C}^n \\ &\psi(\xi) \leqslant 0 && \forall \xi \in W \cap i\mathbb{R}^n \\ &\psi(\xi) \leqslant a && \forall \xi \in i\mathbb{R}^n \end{aligned}$$

and such that, moreover, for a constant $c = c(\psi) > 0$ we have

$$\begin{aligned} &|\psi(\xi) - \psi(\xi')| \leqslant c|\xi - \xi'| && \forall \xi, \xi' \text{ in } \mathbb{C}^n \\ &\psi(\Theta) \geqslant \varphi(\Theta) - \varepsilon \\ &\psi(\xi) = H_K(\xi) + \delta|\mathrm{Re}\,\xi| && \forall \xi \in \mathbb{C}^n - B(\Theta, 2r) \,. \end{aligned}$$

PROOF. α) It is not restrictive to assume that $4\delta < a$. We then choose $\delta_1 > 0$ according to lemma 14 so that given $\Theta \in W$ with $|\mathrm{Im}\,\Theta| = 1$ $|\mathrm{Re}\,\Theta| < r/2$ and given any plurisubharmonic function φ on $B(\Theta, r)$ with

$$\varphi(\xi) \leqslant H_K(\xi) + \delta_1|\xi| \qquad \forall \xi \in B(\Theta, r)$$

$$\varphi(\xi) \leqslant 0 \qquad \forall \xi \in W \cap i\mathbb{R}^n \cap B(\Theta, r)$$

we can find a global plurisubharmonic function g on $\mathbb{C}^n$ such that

$$g(\xi) \leqslant H_K(\xi) + \delta|\xi| \qquad \forall \xi \in \mathbb{C}^n$$

$$g(\xi) \leqslant 0 \qquad \forall \xi \in W \cap i\mathbb{R}^n$$

$$g(\Theta) \geqslant \varphi(\Theta)$$

$$g(\xi) = H_K(\xi) + \delta|\mathrm{Re}\,\xi| \qquad \forall \xi \in \mathbb{C}^n - B(\Theta, r)\,.$$

Let now $\chi \in \mathfrak{D}(\mathbb{C}^n)$ be a C^∞ function with compact support for which we assume

$$\mathrm{supp}\,\chi \subset \{\xi \in \mathbb{C}^n |\ |\xi| < 1\}$$

$$\chi(\xi) \geqslant 0 \qquad \forall \xi \in \mathbb{C}^n$$

$$\int \chi(\xi)\, d\lambda(\xi) = 1 \qquad \text{(where } d\lambda(\xi) \text{ is the Lebesgue measure in } \mathbb{C}^n\text{)}$$

χ is a function of $|\xi_1|, \ldots, |\xi_n|$ only i.e.

$$\chi(\exp(i\Theta_1)\xi_1, \ldots, \exp(i\Theta_n)\xi_n) = \chi(\xi_1, \ldots, \xi_n)$$

$\forall \xi = (\xi_1, \ldots, \xi_n) \in \mathbb{C}^n$ and for any choice of $\Theta_1, \ldots, \Theta_n$ in $\mathbb{R}$.

For any $\sigma > 0$ we define

$$\chi_\sigma(\xi) = \sigma^{-2n}\chi\left(\frac{\xi}{\sigma}\right)$$

so that χ_σ is supported in the ball $\{\xi \in \mathbb{C}^n |\ |\xi| < \sigma\}$. We then consider for every $\sigma > 0$ the function

$$g_\sigma(\xi) = \chi_\sigma * g = \int \chi_\sigma(\xi - w) g(w)\, d\lambda(w)\,.$$

We make use of theorem 2.6.3 of [14]. According to that theorem we have that

$$g_\sigma \text{ is } C^\infty \text{ and plurisubharmonic on } \mathbb{C}^n$$

$$g_\sigma \searrow g \text{ as } \sigma \to 0 \text{ for every } \xi \in \mathbb{C}^n\,.$$

β) For any convex compact set $F \subset \mathbf{C}^n$ the function

$$H_F(\xi) = \sup_{z \in F} \operatorname{Re} \langle \xi, z \rangle$$

is a subadditive function.

Indeed

$$\begin{aligned} H_F(\xi_1 + \xi_2) &= \sup_{z \in F} \operatorname{Re} \langle \xi_1 + \xi_2, z \rangle \\ &\leqslant \sup_{z \in F} \operatorname{Re} \langle \xi_1, z \rangle + \sup_{z \in F} \operatorname{Re} \langle \xi_2, z \rangle \\ &= H_F(\xi_1) + H_F(\xi_2) . \end{aligned}$$

From this we deduce that

$$(*) \qquad |H_F(\xi_1) - H_F(\xi_2)| \leqslant \sup \{H_F(\xi_1 - \xi_2), H_F(\xi_2 - \xi_1)\} ,$$

in fact, setting $\xi = \xi_1 - \xi_2$ we have

$$H_F(\xi + \xi_2) \leqslant H_F(\xi) + H_F(\xi_2)$$

hence

$$H_F(\xi_1) - H_F(\xi_2) \leqslant H_F(\xi_1 - \xi_2)$$

and, similarly, we get

$$H_F(\xi_2) - H_F(\xi_1) \leqslant H_F(\xi_2 - \xi_1) .$$

Now if $F \subset \{\xi \in \mathbf{C}^n \big| \, |\xi| \leqslant C\}$ we do have

$$(**) \qquad |H_F(\xi)| \leqslant C|\xi| .$$

Combining $(*)$ and $(**)$ we obtain

$$|H_F(\xi_1) - H_F(\xi_2)| \leqslant C|\xi_1 - \xi_2|$$

i.e. the function $H_F(\xi)$ is uniformly Lipschitz continuous.

We apply the previous remark to the compact set

$$F = \{\xi \in \mathbf{R}^n | \operatorname{dist}(\xi, K) \leqslant \delta\}$$

so that

$$H_F(\xi) = H_K(\xi) + \delta |\operatorname{Re} \xi| .$$

We then have

$$\chi_\sigma * (H_K(\xi) + \delta|\operatorname{Re}\xi|) = \int_{B(0,1)} \chi(w)\{H_K(\xi - \sigma w) + \delta|\operatorname{Re}(\xi - \sigma w)|\}\, d\lambda(w)$$
$$\leqslant H_K(\xi) + \delta|\operatorname{Re}\xi| + C\sigma \int_{B(0,1)} \chi(w)|w|\, d\lambda(w)$$
$$\leqslant H_K(\xi) + \delta|\operatorname{Re}\xi| + C\sigma$$

where $C = \sup\{|\xi| \,\big|\, \xi \in \mathbb{R}^n \operatorname{dist}(K, \xi) \leqslant \delta\}$.

γ) We define the sets

$$G_\sigma = \{\xi \in \mathbb{C}^n \big| g_\sigma(\xi) < H_K(\xi) + \delta|\xi| + \varepsilon\}$$
$$L_\sigma = \{\xi \in \mathbb{C}^n | g_\sigma(\xi) < \varepsilon\}\ .$$

Since g_σ is C^∞ the sets G_σ and L_σ are open. Moreover if $0 < \sigma_1 < \sigma_2$ we have

$$G_{\sigma_2} \subset G_{\sigma_1} \qquad L_{\sigma_2} \subset L_{\sigma_1}$$

and moreover as $g_\sigma \searrow g$ for $\sigma \to 0^+$ we obtain that

$$\overline{B(\Theta, 3r)} \subset \bigcup_{\sigma>0} G_\sigma \quad \text{ and } W \cap i\mathbb{R}^n \cap \overline{B(\Theta, 3r)} \subset \bigcup_{\sigma>0} L_\sigma\,.$$

The two sets on the left hand side of these inclusions are compact sets.

It follows that there exists $\sigma_0 > 0$ such that for any σ with $0 < \sigma < \sigma_0$ we have

$$\overline{B(\Theta, 3r)} \subset G_\sigma \quad \text{ and } W \cap i\mathbb{R}^n \cap \overline{B(\Theta, 3r)} \subset L_\sigma\,.$$

This means that for $0 < \sigma < \sigma_0$

(i) $g_\sigma(\xi) - \varepsilon < H_K(\xi) + \delta|\xi| \ \ \forall \xi \in B(\Theta, 3r)$

(ii) $g_\sigma(\xi) - \varepsilon < 0 \ \ \forall \xi \in W \cap i\mathbb{R}^n \cap B(\Theta, 3r)$.

Let's take σ with $0 < \sigma < r$ so that for $\xi \in \mathbb{C}^n - B(\Theta, 2r)$ we have

(iii) $g_\sigma(\xi) = \chi_\sigma * (H_K(\xi) + \delta|\operatorname{Re}\xi|) \leqslant H_K(\xi) + \delta|\operatorname{Re}\xi| + C\sigma$.

Now we choose σ with

$$0 < \sigma < \min\left\{\sigma_0, r, \frac{\varepsilon}{C}\right\}$$

and define

$$\psi(\xi) = \max\{g_\sigma(\xi) - \varepsilon, H_K(\xi) + \delta|\mathrm{Re}\,\xi|\}$$

so that ψ is a global plurisubharmonic function on $\mathbb{C}^n$. Moreover outside the ball $B(\Theta, 2r)$ the function ψ equals $H_K(\xi) + \delta|\mathrm{Re}\,\xi|$ which is uniformly continuous in $\mathbb{C}^n$ while the set

$$\{\xi \in \mathbb{C}^n | g_\sigma(\xi) - \varepsilon \geqslant H_K(\xi) + \delta|\mathrm{Re}\,\xi|\}$$

is compact contained in $B(R, 2r)$ and there ψ equals the C^∞ function $g_\sigma(\xi) - \varepsilon$.

It follows that ψ is uniformly Lipschitz continuous on the whole of $\mathbb{C}^n$, i.e. for some $c = c(\psi) > 0$ we have

$$|\psi(\xi) - \psi(\xi')| \leqslant c|\xi - \xi'| .$$

The function ψ verifies moreover the following inequalities

$$\begin{aligned} &\psi(\xi) \leqslant H_K(\xi) + \delta|\xi| && \forall \xi \in \mathbb{C}^n \text{ ((because of (i))} \\ &\psi(\xi) \leqslant 0 && \forall \xi \in W \cap i\mathbb{R}^n \text{ (because of (ii))} \\ &\psi(\xi) = H_K(\xi) + \delta|\mathrm{Re}\,\xi| && \forall \xi \in \mathbb{C}^n - B(\Theta, 2r) \text{ (because of (iii))} . \end{aligned}$$

The condition

$$\psi(\xi) \leqslant a \qquad \forall \xi \in i\mathbb{R}^n$$

is automatic. In fact $\psi(\xi) \leqslant H_K(\xi) + \delta|\mathrm{Re}\,\xi| = 0$ for $\xi \in i\mathbb{R}^n$ and $\xi \notin B(\Theta, 2r)$. For $\xi \in i\mathbb{R}^n \cap B(\Theta, 2r)$ we do have

$$\begin{aligned} \psi(\xi) &\leqslant \delta|\xi| < \delta\left(1 + \frac{r}{2} + 2r\right) \\ &\leqslant 4\delta \leqslant a . \end{aligned}$$

Finally we have

$$\begin{aligned} \psi(\Theta) &\geqslant g_\sigma(\Theta) - \varepsilon \geqslant g(\Theta) - \varepsilon \\ &\geqslant \varphi(\Theta) - \varepsilon . \end{aligned}$$

This completes the proof.

PROPOSITION 20. *Let $K \subset K'$ compact convex sets in $\mathbb{R}^n$ with $K \subset \mathring{K}'$.*

The Phragmén-Lindelöf principle (I) *on* W *is equivalent to the following statement*

(VI) « *There exist a* $\delta > 0$ *and* $a > 0$ *such that for any choice of a plurisubharmonic function* φ *defined on* $\mathbb{C}^n$ *and satisfying*

(1) $$\varphi(\xi) \leqslant H_K(\xi) + \delta|\xi| \qquad \forall \xi \in \mathbb{C}^n$$

(2) $$\varphi(\xi) \leqslant 0 \qquad \forall \xi \in W \cap i\mathbb{R}^n$$

(3) *for some constant* $c = c(\varphi) > 0$ *we have*

$$|\varphi(\xi) - \varphi(\xi')| \leqslant c|\xi - \xi'| \qquad \forall \xi, \xi' \text{ in } \mathbb{C}^n$$

(4) $$\varphi(\xi) \leqslant a \qquad \forall \xi \in i\mathbb{R}^n$$

we have also that

(5) $$\varphi(\xi) \leqslant H_{K'}(\xi) \qquad \forall \xi \in W \text{ »}.$$

PROOF. We will prove that (I) $\Rightarrow$ (VI) $\Rightarrow$ (IV). By the equivalence of (I) and (IV) the proposition will be proved.

The implication (I) $\Rightarrow$ (VI) is straightforward.

Let now $\delta > 0$ $a > 0$ be such that condition (VI) holds. We fix r with $0 < r < 1$ and choose $\delta_1 > 0$ so that lemma 15 holds. Let $\Theta \in W$ with $|\operatorname{Im} \Theta| = 1$ and let φ be a plurisubharmonic function defined on $B(\Theta, r)$ and such that

$$\varphi(\xi) \leqslant H_K(\xi) + \delta_1|\xi| \qquad \forall \xi \in B(\Theta, r)$$

$$\varphi(\xi) \leqslant 0 \qquad \forall \xi \in W \cap i\mathbb{R}^n \cap B(\Theta, r)\,.$$

If $|\operatorname{Re} \Theta| \geqslant r/2$ we can argue as in proposition 19 and conclude that $\varphi(\Theta) \leqslant H_{K'}(\Theta)$ provided δ_1 is chosen sufficiently small.

We can therefore assume $|\operatorname{Re} \Theta| < r/2$ and we can apply lemma 15. For any choice of $\varepsilon > 0$ we can construct a plurisubharmonic function ψ on $\mathbb{C}^n$ such that

$$\psi(\xi) \leqslant H_K(\xi) + \delta|\xi| \qquad \forall \xi \in \mathbb{C}^n$$

$$\psi(\xi) \leqslant 0 \qquad \forall \xi \in W \cap i\mathbb{R}^n$$

$$\psi(\xi) \leqslant a \qquad \forall \xi \in i\mathbb{R}^n$$

for some constant $c(\psi) = c > 0$ we have

$$|\psi(\xi) - \psi(\xi')| \leqslant c|\xi - \xi'|$$
$$\psi(\Theta) \geqslant \varphi(\Theta) - \varepsilon .$$

Because condition (VI) holds we deduce that

$$\psi(\Theta) \leqslant H_{K'}(\Theta)$$

and therefore

$$\varphi(\Theta) - \varepsilon \leqslant H_{K'}(\Theta) .$$

As $\varepsilon > 0$ is arbitrary we must have

$$\varphi(\Theta) \leqslant H_{K'}(\Theta) .$$

This shows that condition (IV) also holds.

PROPOSITION 21. *Let $K \subset K'$ be compact convex sets in $\mathbb{R}^n$ with $K \subset \mathring{K}'$. The Phragmén-Lindelöf principle* (I) *on W is equivalent to the following statement*

(VII) « *There is a $\delta > 0$ such that for any choice of a plurisubharmonic function φ on $\mathbb{C}^n$ satisfying*

(1) $$\varphi(\xi) \leqslant H_K(\xi) + \delta|\xi| \qquad \forall \xi \in \mathbb{C}^n$$

(2) $$\varphi(\xi) \leqslant 0 \qquad \forall \xi \in W \cap i\mathbb{R}^n$$

(3) *for some constant $c = c(\varphi) > 0$ we have*

$$|\varphi(\xi) - \varphi(\xi')| \leqslant c|\xi - \xi'| \qquad \forall \xi, \xi' \text{ in } \mathbb{C}^n$$

we also have

(4) $$\varphi(\xi) \leqslant H_{K'}(\xi) \qquad \forall \xi \in W \text{ ».}$$

PROOF. We have the obvious implications

$$(\text{V}) \Rightarrow (\text{VII}) \Rightarrow (\text{VI}) .$$

Since (V) and (VI) are equivalent to (I) we have the desired conclusion.

PROPOSITION 22. *Let $K \subset K' \subset K''$ be compact convex sets in $\mathbb{R}^n$ with*

$$K \subset \mathring{K}', \qquad K' \subset \mathring{K}''.$$

The Phragmén-Lindelöf principle (I) *on W is equivalent to the following statement*

(VIII) « *We can find real numbers $\delta > 0$ and $a > 0$ such that for any choice of a plurisubharmonic function φ on $\mathbb{C}^n$ satisfying*

(1) $$\varphi(\xi) \leqslant H_K(\xi) + \delta|\xi| \qquad \forall \xi \in \mathbb{C}^n$$

(2) $$\varphi(\xi) \leqslant H_{K''}(\xi) \qquad \forall \xi \in W$$

(3) $$\varphi(\xi) \leqslant a \qquad \forall \xi \in i\mathbb{R}^n$$

(4) *for some constant $c = c(\varphi) > 0$ we have*

$$|\varphi(\xi) - \varphi(\xi')| \leqslant c|\xi - \xi'| \qquad \forall \xi, \xi' \text{ in } \mathbb{C}^n$$

we have also that

(5) $$\varphi(\xi) \leqslant H_{K'}(\xi) \qquad \forall \xi \in W \text{ ».}$$

PROOF. We will show that (V) $\Rightarrow$ (VIII) $\Rightarrow$ (IV). This will suffice since (V) and (IV) are equivalent to (I).

Note that if $\xi \in W \cap i\mathbb{R}^n$ from condition (2) above we deduce $\varphi(\xi) \leqslant 0$.

This shows that (V) $\Rightarrow$ (VIII).

We want now to show that (VIII) $\Rightarrow$ (IV).

We first remark that in condition (VIII) it is not restrictive to assume that K contains the origin of the coordinates. In fact let K be compact contained in $\mathbb{R}^n$ and let $z_0 \in K$. Set $K_1 = K - z_0$ so that $0 \in K_1$.

We have $\forall \xi \in \mathbb{C}^n$

$$\begin{aligned} H_{K_1}(\xi) &= \sup_{z \in K_1} \operatorname{Re}\langle \xi, z\rangle = \sup_{z \in K} \operatorname{Re}\langle \xi, z\rangle - \operatorname{Re}\langle \xi, z_0\rangle \\ &= H_K(\xi) - \operatorname{Re}\langle \xi, z_0\rangle . \end{aligned}$$

Replacing in condition (VIII) K, K', K'' with $K_1 = K - z_0$, $K_1' = K' - z_0$, $K_1'' = K'' - z_0$ and φ by $\varphi - \operatorname{Re}\langle \xi, z_0\rangle = \varphi_1$ we see that φ_1 satisfies conditions (1) (2) (3) (5) with K_1, K_1', K_1''. Moreover φ_1 satisfies condition (4) because $\operatorname{Re}\langle \xi, z_0\rangle$ is uniformly Lipschitz continuous on all of $\mathbb{C}^n$.

Conversely if condition (VIII) is satisfied with φ_1 and K_1, K_1', K_1'' with $0 \in K_1$, given $z_0 \in \mathbb{R}^n$ replacing K_1, K_1', K'' with $K = K_1 + z_0$, $K' = K_1' + z_0$,

$K'' = K_1'' + z_0$ and φ_1 by $\varphi = \varphi_1 + \operatorname{Re}\langle \xi, z_0 \rangle$ we have all conditions of (VIII) satisfied again.

We therefore assume $0 \in K$ and that (VIII) holds.

We claim that if $\varrho > 1$ is so chosen that $\varrho K' \subset K''$ and if φ is a plurisubharmonic function on $\mathbb{C}^n$ which satisfies conditions (1), (3) and (4) and for some integer $m \geqslant 1$ the condition

$$\varphi(\xi) \leqslant H_{\varrho^m K''}(\xi) = \varrho^m H_{K''}(\xi) \qquad \forall \xi \in W$$

(instead of condition (2)), then the same function φ satisfies also condition (2)

$$\varphi(\xi) \leqslant H_{K''}(\xi) \qquad \forall \xi \in W$$

and therefore also the conclusive condition (5)

$$\varphi(\xi) \leqslant H_{K'}(\xi) \qquad \forall \xi \in W\,.$$

In fact $\varrho^{-m}\varphi(\xi)$ satisfies conditions (1), (2), (3) and (4) and therefore also (5)

$$\varrho^{-m}\varphi(\xi) \leqslant H_{K'}(\xi) \qquad \forall \xi \in W\,.$$

This implies

$$\varrho^{-m+1}\varphi(\xi) \leqslant \varrho H_{K'}(\xi) = H_{\varrho K'}(\xi) \leqslant H_{K''}(\xi)\,.$$

So that $\varrho^{-m+1}\varphi(\xi)$ satisfies condition (1), (2), (3) and (4) and thus also (5)

$$\varrho^{-m+1}\varphi(\xi) \leqslant H_{K'}(\xi) \qquad \forall \xi \in W\,.$$

Arguing as before by a finite number of steps we deduce that

$$\varphi(\xi) \leqslant H_{K''}(\xi) \qquad \forall \xi \in W$$

as we wanted and therefore also

$$\varphi(\xi) \leqslant H_{K'}(\xi) \qquad \forall \xi \in W\,.$$

Now given $\delta > 0$, $a > 0$ and r with $0 < r < 1$ we choose $\delta_1 > 0$ as prescribed by lemma 15. Let $\Theta \in W$ with $|\operatorname{Im}\Theta| = 1$ and let φ be a plurisubharmonic function defined on $B(\Theta, r)$ such that

$$\begin{aligned} &\varphi(\xi) \leqslant H_K(\xi) + \delta_1|\xi| && \forall \xi \in B(\Theta, r) \\ &\varphi(\xi) \leqslant 0 && \forall \xi \in W \cap i\mathbb{R}^n \cap B(\Theta, r)\,. \end{aligned}$$

We want to show that

$$\varphi(\Theta) \leqslant H_{K'}(\Theta) .$$

If $|\mathrm{Re}\,\Theta| \geqslant r/2$ as in proposition 19 we deduce the inequality $\varphi(\Theta) \leqslant H_{K'}(\Theta)$ from the inequality $\varphi(\xi) \leqslant H_K(\xi) + \delta_1|\xi|$ provided $\delta_1 > 0$ has been chosen sufficiently small.

We can therefore assume $|\mathrm{Re}\,\Theta| < r/2$ and we can apply lemma 15.

Given $\varepsilon > 0$ we can construct a plurisubharmonic function ψ on $\mathbf{C}^n$ with the properties

$$\psi(\xi) \leqslant H_K(\xi) + \delta|\xi| \qquad \forall \xi \in \mathbf{C}^n$$

$$\psi(\xi) \leqslant 0 \qquad \forall \xi \in W \cap i\mathbf{R}^n$$

there exists a constant $c = c(\psi) > 0$ such that

$$|\psi(\xi) - \psi(\xi')| \leqslant c|\xi - \xi'| \quad \text{for } \xi, \xi' \text{ in } \mathbf{C}^n$$

$$\psi(\xi) \leqslant a \qquad \forall \xi \in i\mathbf{R}^n$$

$$\psi(\Theta) \geqslant \varphi(\Theta) - \varepsilon .$$

Moreover we have

$$\psi(\xi) = H_K(\xi) + \delta|\mathrm{Re}\,\xi| \qquad \forall \xi \in \mathbf{C}^n - B(\Theta, 2r) .$$

We claim that for some large m we must have

$$(*) \qquad \psi(\xi) - \varepsilon \leqslant \varrho^m H_{K''}(\xi) \qquad \forall \xi \in W .$$

Assume, if possible, that this is not true and choose for every m a point $\xi_m \in W$ such that

$$\psi(\xi_m) - \varepsilon > \varrho^m H_{K''}(\xi_m) .$$

For large m we have

$$H_K(\xi) + \delta|\mathrm{Re}\,\xi| \leqslant \varrho^m H_{K''}(\xi) \qquad \forall \xi \in \mathbf{C}^n ;$$

in fact if $K(\delta)$ denotes the δ-neighborhood of K in $\mathbf{R}^n$ the above inequality is equivalent to the inequality

$$H_{K(\delta)}(\xi) \leqslant H_{\varrho^m K''}(\xi)$$

which is true as soon as $K(\delta) \subset \varrho^m K''$.

This shows that for large m, $\xi_m \in B(\Theta, 2r)$.

Moreover let

$$M = \sup_{\xi \in B(\Theta, 2r)} (\psi(\xi) - \varepsilon)$$

and let $\sigma > 0$ be such that $\{x \in \mathbb{R}^n \big| \, |x| < \sigma\} \subset K''$; then

$$H_{K''}(\xi) = \sup_{x \in K''} \operatorname{Re} \langle \xi, x \rangle \geqslant \sigma |\operatorname{Re} \xi| .$$

Therefore

$$\sigma |\operatorname{Re} \xi_m| < M \varrho^{-m} .$$

This shows that, for $m \to \infty$ $\operatorname{Re} \xi_m \to 0$.

By passing to a subsequence we may assume that ξ_m converges to some point $\xi^{(0)} \in B(\Theta, 2r)$. We must have $\xi^{(0)} \in W$ and $\operatorname{Re} \xi^{(0)} = 0$ and also $\psi(\xi^{(0)}) - \varepsilon \geqslant 0$ i.e. $\psi(\xi^{(0)}) \geqslant \varepsilon > 0$.

This is a contradiction because $\operatorname{Re} \xi^{(0)} = 0$ so that $\xi^{(0)} \in W \cap i\mathbb{R}^n$ and there we have $\psi(\xi^{(0)}) \leqslant 0$.

We conclude then that, for large m, $(*)$ holds. By what we have established at the beginning we must have (since (VIII) holds) that $\psi(\xi) - \varepsilon \leqslant H_{K''}(\xi)$, $\forall \xi \in W$, and consequently also

$$\psi(\xi) - \varepsilon \leqslant H_{K'}(\xi) \qquad \forall \xi \in W .$$

In particular for $\xi = \Theta$ we get

$$\varphi(\Theta) \leqslant \psi(\Theta) + \varepsilon \leqslant H_{K'}(\Theta) + 2\varepsilon .$$

As $\varepsilon > 0$ is arbitrary we deduce that

$$\varphi(\Theta) \leqslant H_{K'}(\Theta) .$$

Therefore condition (IV) holds (with δ_1 in place of δ).

PROPOSITION 23. *Let $K \subset K' \subset K''$ be compact convex sets in $\mathbb{R}^n$ with*

$$K \subset \mathring{K}' , \qquad K' \subset \mathring{K}'' .$$

The Phragmén-Lindelöf principle (I) *on* W *is equivalent to the following statement*

(IX) « *We can find real numbers $\delta > 0$ and $a > 0$ such that for every* *integer* $N > 0$ *and for any choice of an entire function* $F(\xi)$ *on* $\mathbb{C}^n$ *satisfying the conditions*

$$\text{(1)} \qquad \frac{1}{N}\log|F(\xi)| \leqslant H_K(\xi) + \delta|\xi| \qquad \forall \xi \in \mathbb{C}^n$$

$$\text{(2)} \qquad \frac{1}{N}\log|F(\xi)| \leqslant H_{K''}(\xi) \qquad \forall \xi \in W$$

$$\text{(3)} \qquad \frac{1}{N}\log|F(\xi)| \leqslant a \qquad \forall \xi \in i\mathbb{R}^n$$

(4) *for every integer* $k \geqslant 0$ *the function*

$$(1 + |\xi|)^k |F(\xi)| \exp\big(-N\big(H_K(\xi) + \delta|\mathrm{Re}\,\xi|\big)\big)$$

is bounded on $\mathbb{C}^n$

we have also that

$$\text{(5)} \qquad \frac{1}{N}\log|F(\xi)| \leqslant H_{K'}(\xi) \qquad \forall \xi \in W \text{ ».}$$

PROOF. α) We first show that (V) $\Rightarrow$ (IX) (and therefore that (I) $\Rightarrow$ (IX)).

In fact if $F(\xi)$ is an entire function which satisfies conditions (1) and (2) of (IX) for some integer $N > 0$, we have that $\varphi(\xi) = 1/N \log|F(\xi)|$ is a plurisubharmonic function on $\mathbb{C}^n$ for which

$$\varphi(\xi) \leqslant H_K(\xi) + \delta|\xi| \qquad \forall \xi \in \mathbb{C}^n$$

$$\varphi(\xi) \leqslant 0 \qquad \forall \xi \in W \cap i\mathbb{R}^n$$

(because for $\xi \in i\mathbb{R}^n$ $H_{K''}(\xi) = 0$). By condition (V) we deduce then that

$$\varphi(\xi) \leqslant H_{K'}(\xi) \qquad \forall \xi \in W$$

i.e. condition (5) of (IX). Thus (IX) holds if (V) holds.

β) Conversely we will show that if (IX) holds with K, K', K'', and 2δ instead of δ we will have that (VIII) holds with K, K' and K''' instead of K'' with

$$K' \subset \mathring{K}''' , \qquad K''' \subset \mathring{K}''$$

and with $\delta > 0$ so small that

$$H_{K'''}(\xi) + \delta|\mathrm{Re}\,\xi| \leqslant H_{K''}(\xi) \qquad \forall \xi \in \mathbb{C}^n .$$

Let φ be a plurisubharmonic function on $\mathbb{C}^n$ which satisfies conditions (1), (2), (3), (4) of statement (VIII) (with K''' in place of K'').

Let $N=1$ be an integer and consider the function

$$N\varphi\left(\frac{\xi}{N}\right).$$

If $c = c(\varphi)$ is the Lipschitz constant for φ then the new function $N\varphi(\xi/N)$ is uniformly Lipschitz continuous with the same Lipschitz constant.

Indeed

$$\left|N\varphi\left(\frac{\xi}{N}\right) - N\varphi\left(\frac{\xi'}{N}\right)\right| \leqslant Nc\left|\frac{\xi}{N} - \frac{\xi'}{N}\right|$$
$$= c|\xi - \xi'|\,.$$

Now we make use of the following theorem of Hörmander ([14] theorem 4.4.3) « Let ψ be a plurisubharmonic function in $\mathbb{C}^n$ such that for some constant c

$$|\psi(\xi) - \psi(\xi')| < c \quad \text{if } |\xi - \xi'| < 1\,.$$

Let Σ be a complex affine linear subspace of $\mathbb{C}^n$ of codimension k. For every analytic function u in Σ such that

$$\int_{\Sigma} |u|^2 \exp(-\psi)\, d\sigma < \infty$$

where $d\sigma$ denotes the Lebesgue measure in Σ, there exists an analytic function U in $\mathbb{C}^n$ such that $U = u$ in Σ and

$$\int_{\mathbb{C}^n} |U|^2 \exp(-\psi)(1 + |\xi|^2)^{-3k}\, d\lambda \leqslant 6^k \pi^k \exp(kc) \int_{\Sigma} |u|^2 \exp(-\psi)\, d\sigma$$

where $d\lambda$ is the Lebesgue measure in $\mathbb{C}^n$ ».

Let $\Theta \in W$, we apply the previous theorem with $\Sigma = \{N\Theta\}$ and with $\psi(\xi) = 2N\varphi(\xi/N)$. We deduce that there exists an entire function $f_N(\xi)$ in $\mathbb{C}^n$ with

$$f_N(N\Theta) = \exp\left(N\varphi(\Theta)\right)$$

and such that

$$(*) \qquad \int_{\mathbb{C}^n} |f_N(\xi)|^2 (1 + |\xi|^2)^{-3n} \exp\left(-2N\varphi\left(\frac{\xi}{N}\right)\right) d\lambda(\xi) \leqslant \left(6\pi \exp(2c)\right)^n$$

since

$$|f_N(N\theta)|^2 \exp\left(-2N\varphi\left(\frac{N\Theta}{N}\right)\right) = 1 .$$

γ) We want to deduce from the above estimate $(*)$ a pointwise estimate.

For that purpose we use the following

LEMMA. *Let $\psi: \mathbb{C}^n \to \mathbb{R}$ be uniformly Lipschitz continuous on $\mathbb{C}^n$:*

$$|\psi(\xi) - \psi(\xi')| \leqslant c|\xi - \xi'| \qquad \text{for } \xi, \xi' \text{ in } \mathbb{C}^n .$$

Let g be an entire function on $\mathbb{C}^n$ such that $ge^{-\psi}$ is square integrable:

$$\int_{\mathbb{C}^n} |g(\xi)|^2 \exp(-2\psi(\xi))\, d\lambda(\xi) < \infty .$$

Then for every $\xi \in \mathbb{C}^n$ we have

$$|g(\xi) \exp(-\psi(\xi))| \leqslant \sqrt{\frac{n!}{\pi^n}} \exp(c) \int_{\mathbb{C}^n} |g(w)|^2 \exp(-2\psi(w))\, d\lambda(w) .$$

Note that $\pi^n/n!$ is the (Lebesgue) measure of the unit ball $B(\xi, 1)$ in $\mathbb{C}^n$.

Since g is holomorphic it is in particular a harmonic function so that by the mean value theorem we have for every $\xi \in \mathbb{C}^n$

$$g(\xi) = \frac{n!}{\pi^n} \int_{B(\xi,1)} g(w)\, d\lambda(w) .$$

From this equality we derive

$$g(\xi) \exp(-\psi(\xi)) = \frac{n!}{\pi^n} \int_{B(\xi,1)} g(w) \exp(-\psi(\xi))\, d\lambda(w) .$$

Now on the ball $B(\xi, 1)$ we have

$$\exp(-\psi(\xi)) = \exp(-\psi(w)) \cdot \exp(\psi(w) - \psi(\xi)) \leqslant \exp(c) \cdot \exp(-\psi(w))$$

where c is the Lipschitz constant of ψ.

Therefore

$$|g(\xi) \exp(-\psi(\xi))| \leqslant \exp(c) \frac{n!}{\pi^n} \int_{B(\xi,1)} |g(w)| \exp(-\psi(w))\, d\lambda(w) .$$

By Schwarz inequality

$$\int\limits_{B(\xi,1)} |g(w)| \exp(-\psi(w))\, d\lambda(w) \leqslant \Bigl(\int\limits_{B(\xi,1)} d\lambda(w)\Bigr)^{\frac{1}{2}} \Bigl(\int\limits_{B(\xi,1)} |g(w)|^2 \exp(-2\psi(w))\, d\lambda(w)\Bigr)^{\frac{1}{2}}$$

$$\leqslant \sqrt{\frac{\pi^n}{n!}} \Bigl(\int\limits_{\mathbb{C}^n} |g(w)|^2 \exp(-2\psi(w))\, d\lambda(w)\Bigr)^{\frac{1}{2}}.$$

Combining this last inequality with the previous one we obtain the statement of the lemma.

We now apply the previous lemma for $g(\xi) = f_N(\xi)$ and for $\psi(\xi) = N\varphi(\xi/N) + (3n/2) \log(1 + |\xi|^2)$.

This we can do, since $\log(1 + t^2)$ is uniformly Lipschitz with Lipschitz constant 1, so that $\psi(\xi)$ is uniformly Lipschitz on $\mathbb{C}^n$ with Lipschitz constant $c + 3n/2$ $(c = c(\varphi))$.

We obtain therefore the pointwise estimate from $(*)$

$$(**) \qquad |f_N(\xi)| \leqslant c'(1 + |\xi|^2)^{3n/2} \exp N\varphi\left(\frac{\xi}{N}\right)$$

where

$$c' = (6\pi \exp[2c])^{n/2} \exp\left(c + \frac{3n}{2}\right) \sqrt{\frac{n!}{\pi^n}}$$

and $c = c(\varphi)$.

δ) We now choose an entire function $g(\xi)$ on $\mathbb{C}^n$ with the following properties

$$(a) \qquad g(\Theta) \neq 0$$

$$(b) \qquad c'(1 + |\xi|^2)^{3n/2} |g(\xi)| \leqslant \exp(\delta |\mathrm{Re}\, \xi|) \qquad \forall \xi \in \mathbb{C}^n$$

(c) for every integer $k \geqslant 0$ the function

$|g(\xi)|(1 + |\xi|^2)^{k/2} \exp(-\delta |\mathrm{Re}\, \xi|)$

is bounded on $\mathbb{C}^n$.

For this purpose we take a C^∞ function χ on $\mathbb{R}^n$ with support contained in the ball $\{x \in \mathbb{R}^n \mid |x| < \delta\}$ and such that

$$\int\limits_{\mathbb{R}^n} \chi(x) \exp\langle\Theta, x\rangle\, dx \neq 0$$

where $\langle \xi, x\rangle = \sum_{i=1}^{n} \xi_i x_i$. We consider then the Laplace transform of χ

$$\tilde{\chi}(\xi) = \int_{\mathbf{R}^n} \chi(x) \exp(\langle \xi, x\rangle)\, dx .$$

We have $\tilde{\chi}(\Theta) \neq 0$. Moreover from the necessity part of Paley-Wiener theorem (the easy part) we deduce that, for every integer $k \geqslant 0$ we can find a constant $c_k > 0$ such that

$$|\tilde{\chi}(\xi)| \leqslant c_k(1 + |\xi|^2)^{-k/2} \exp(\delta |\mathrm{Re}\, \xi|) \qquad \forall \xi \in \mathbf{C}^n .$$

The assumptions (a) (b) (c) will be satisfied if we set

$$g(\xi) = c''\, \tilde{\chi}(\xi)$$

with $c'' > 0$ and $c'' c_{3n} c' < 1$.

We now set by definition

$$F_N(\xi) = f_N(N\xi) g(\xi) .$$

We claim that for this entire function conditions (1) (2) (3) and (4) of statement (IX) are satisfied with 2δ in place of δ.

For condition (1) we have

$$(***) \quad \frac{1}{N} \log |F_N(\xi)| = \frac{1}{N} \log \left| F_N(N\xi) c'^{-1} (1 + |\xi|^2)^{-3n/2} \right| + \frac{1}{N} \log \left| c'(1 + |\xi|^2)^{3n/2} g(\xi) \right| \leqslant \varphi(\xi) + \frac{\delta}{N} |\mathrm{Re}\, \xi|$$

because of the inequalities $(**)$ and (b). Since $\varphi(\xi) \leqslant H_K(\xi) + \delta|\xi|$ $\forall \xi \in \mathbf{C}^n$ we deduce

$$\frac{1}{N} \log |F_N(\xi)| \leqslant H_K(\xi) + 2\delta |\xi| \qquad \forall \xi \in \mathbf{C}^n .$$

For condition (2) we argue as follows. By assumption on the function φ we have (condition (2) of VIII with K''' instead of K'')

$$\varphi(\xi) \leqslant H_{K'''}(\xi) \qquad \forall \xi \in W .$$

From (∗∗∗) we derive then

$$\frac{1}{N}\log|F_N(\xi)|\leqslant H_{K'''}(\xi)+\delta|\mathrm{Re}\,\xi|\qquad\forall\xi\in W$$
$$\leqslant H_{K''}(\xi)$$

because of the assumptions we have made above about K'', K''' and δ. For condition (3) the argument is the following.

For $\xi\in i\mathbb{R}^n$ we get from (∗∗∗)

$$\frac{1}{N}\log|F_N(\xi)|\leqslant\varphi(\xi)\leqslant a$$

the last inequality being a consequence of the hypothesis that φ satisfies condition (3) of (VIII).

It remains to verify condition (4) (with 2δ instead of δ) which is equivalent to the statement that

$$(1+|\xi|^2)^{k/2}|F_N(\xi)|\exp\big(-N(H_K(\xi)+2\delta|\mathrm{Re}\,\xi|)\big)$$

is a bounded function on $\mathbb{C}^n$.

Now we have proved that (see (∗∗))

$$|f_N(\xi)|(1+|\xi|^2)^{-3n/2}\exp\left(-N\varphi\left(\frac{\xi}{N}\right)\right)$$

is bounded on $\mathbb{C}^n$.

On the other hand the plurisubharmonic function φ on $\mathbb{C}^n$ satisfies by hypothesis the two estimates

$$\varphi(\xi)\leqslant H_K(\xi)+\delta|\xi|\qquad\forall\xi\in\mathbb{C}^n$$
$$\varphi(\xi)\leqslant a\qquad\forall\xi\in i\mathbb{R}^n\,.$$

By the classical Phragmén-Lindelöf theorem for $\mathbb{C}^n$ we deduce that

$$\varphi(\xi)\leqslant a+H_K(\xi)+\delta|\mathrm{Re}\,\xi|\qquad\forall\xi\in\mathbb{C}^n\,.$$

In particular we must have, a fortiori, that the function

$$|f_N(\xi)|(1+|\xi|^2)^{-3n/2}\exp\big(-(H_K(\xi)+\delta|\mathrm{Re}\,\xi|)\big)$$

is bounded on $\mathbb{C}^n$.

Now we have

$$F_N(\xi) = f_N(N\xi)g(\xi)$$

and we note that for every integer N we can find positive constants $c_N > 0$, $c'_N > 0$, such that

$$c'_N \leqslant \frac{(1 + |N\xi|^2)^{3n/2}}{(1 + |\xi|^2)^{3n/2}} \leqslant c_N .$$

Therefore

$$|f_N(N\xi)|(1 + |\xi|^2)^{-3n/2} \exp\big(-N(H_K(\xi) + \delta|\mathrm{Re}\,\xi|)\big)$$

is bounded on $\mathbb{C}^n$.

Our contention will be proved if we show that for every integer $k \geqslant 0$

$$|g(\xi)|(1 + |\xi|^2)^{(k+3n)/2} \exp(-N\delta|\mathrm{Re}\,\xi|)$$

is a bounded function on $\mathbb{C}^n$. But this is exactly property (c) (with $k + 3n$ replacing k). We conclude therefore that also condition (4) of (IX) holds.

Consequently, since we assume that statement (IX) holds we get

$$\frac{1}{N}\log|F_N(\Theta)| \leqslant H_{K'}(\Theta)$$

i.e.

$$\varphi(\Theta) + \frac{1}{N}\log|g(\Theta)| \leqslant H_{K'}(\Theta) .$$

This being true for every integer $N \geqslant 1$ since g is independent from N and $g(\Theta) \neq 0$ we derive

$$\varphi(\Theta) \leqslant H_{K'}(\Theta) .$$

This will be true for any $\Theta \in W$ i.e.

$$\varphi(\xi) \leqslant H_{K'}(\xi) \qquad \forall \xi \in W .$$

This shows that condition (VIII) holds as a consequence of statement (IX), as we claimed.

SECTION 15

Proof of Theorem 3

Let $A_0(D): \mathcal{E}^{p_0}(\Omega) \to \mathcal{E}^{p_1}(\Omega)$ be a differential operator with constant coefficients defined on $\mathbb{R}^n$ (Ω open set of $\mathbb{R}^n$) for which we consider a corresponding Hilbert complex (1) as in section 8. We resume the notations of that section. Let $\mathcal{T} = \mathbb{C}[\xi_1, \dots, \xi_n]$ and let us consider the $\mathcal{T}$-module

$$\operatorname{Im}\left\{\mathcal{T}^{p_1} \xrightarrow{{}^tA_0(\xi)} \mathcal{T}^{p_0}\right\}.$$

Of this module we consider a primary decomposition, the corresponding prime ideals $\mathfrak{p}_1, \dots, \mathfrak{p}_k$, the associated algebraic varieties

$$V_i = \{\xi \in \mathbb{C}^n | p(\xi) = 0 \ \ \forall p \in \mathfrak{p}_i\}, \qquad 1 \leqslant i \leqslant k,$$

and for each V_i the corresponding asymptotic variety $W = V_i^0$ that we decompose into irreducible components. Let W_j $1 \leqslant j \leqslant l$ be the irreducible cones that thus appear when i varies between 1 and k.

With these notations we have the statement of theorem 3. Of that theorem the sufficiency of the condition has already been established, we wish to prove now its necessity.

We have already remarked in section 13 that it is enough to establish the necessity part of theorem 3 under the additional assumption that we now make, that the module

$$\operatorname{Im}\left\{\mathcal{T}^{p_1} \xrightarrow{{}^tA_0(\xi)} \mathcal{T}^{p_0}\right\}$$

is a primary module so that now $k = 1$; there is only one algebraic variety V ($= V_i$) and thus only one asymptotic variety W which decomposes into irreducible components all of the same dimension.

Due to the equivalent formulations of Phragmén-Lindelöf principle (I) (I′) we can thus state our aim as follows

Let ω be an open convex set in $\mathbb{R}^n$ and assume that

$$H^1(\omega, \mathcal{A}_{A_0}) = 0.$$

Then for any compact convex set $K \subset \omega$ we can find $\delta > 0$ and a compact convex set K' with $K \subset K' \subset \omega$ such that the Phragmén-Lindelöf principle (I) *holds for the asymptotic variety W.*

Due to the equivalence of statements (I) and (IX) for W the statement we need to prove can be formulated equivalently as follows:

Let ω be an open and convex set in $\mathbb{R}^n$ and assume that

$$H^1(\omega, \mathcal{A}_{A_0}) = 0 .$$

Then for any compact convex set $K \subset \omega$ we can find $\delta > 0$, a compact convex set K' with $K \subset \mathring{K}' \subset K' \subset \omega$, such that for any choice of $a > 0$, for any choice of a compact convex K'' with $K' \subset K''$, for every integer $N > 0$ and for any choice of an entire function $F(\xi)$ in $\mathbb{C}^n$ satisfying the conditions

(1) $$\frac{1}{N} \log |F(\xi)| \leqslant H_K(\xi) + \delta|\xi| \qquad \forall \xi \in \mathbb{C}^n$$

(2) $$\frac{1}{N} \log |F(\xi)| \leqslant H_{K''}(\xi) \qquad \forall \xi \in W$$

(3) $$\frac{1}{N} \log |F(\xi)| \leqslant a \qquad \forall \xi \in i\mathbb{R}^n$$

(4) *for every integer $k \geqslant 0$ the function*

$$(1 + |\xi|)^k |F(\xi)| \exp\big(-N(H_K(\xi) + \delta|\mathrm{Re}\,\xi|)\big)$$

is bounded on $\mathbb{C}^n$

we have also that

(5) $$\frac{1}{N} \log |F(\xi)| \leqslant H_{K'}(\xi) \qquad \forall \xi \in W .$$

b) Let $\mathfrak{L}(\xi, D_\xi)$ be the noetherian operator associated to the primary module $\mathrm{Im}\left\{\mathcal{P}^{p_1} \xrightarrow{{}^tA_0(\xi)} \mathcal{P}^{p_0}\right\}$. For an entire function $G\colon \mathbb{C}^n \to \mathbb{C}^{p_0}$ we are interested in the values of

$$\mathfrak{L}(\xi, D_\xi) G(\xi) \quad \text{for } \xi \in V .$$

First of all we can find integers m_1, m_2 non negative and a constant $c > 0$ such that for any choice of G we have

$$|\mathfrak{L}(\xi, D_\xi) G(\xi)| \leqslant c(1 + |\xi|)^{m_1} \sum_{|\alpha| \leqslant m_2} |D_\xi^\alpha G(\xi)|$$

for every $\xi \in \mathbb{C}^n$. This because $\mathcal{L}$ is a differential operator with polynomial coefficients.

From Cauchy integral formula we derive for every multindex α the existence of a constant $c_\alpha > 0$ such that, for every G we have

$$|D_\xi^\alpha G(\xi)| \leqslant c_\alpha \sup_{|w-\xi|\leqslant 1} |G(w)| \qquad \forall \xi \in \mathbb{C}^n .$$

Therefore there exists a constant $c' > 0$ such that

$$|\mathcal{L}(\xi, D_\xi) G(\xi)| \leqslant c'(1 + |\xi|)^{m_1} \sup_{|w-\xi|\leqslant 1} |G(w)| \qquad \forall \xi \in \mathbb{C}^n .$$

Let now $\psi\colon \mathbb{C}^n \to \mathbb{R}$ be a uniformly Lipschitz continuous function:

$$|\psi(\xi) - \psi(\xi')| \leqslant \gamma |\xi - \xi'| \qquad \forall \xi, \xi' \text{ in } \mathbb{C}^n ,$$

($\gamma > 0$ is the Lipschitz constant). We note also that $(1 + |\xi|)^{m_1} = \exp\big(m_1 \log(1 + |\xi|)\big)$ and that $\log(1 + t)$ for $t \geqslant 0$ is uniformly Lipschitz continuous with Lipschitz constant 1. We derive then the inequality

$$\begin{aligned} |\mathcal{L}(\xi, D_\xi) G(\xi)| \exp(-\psi(\xi)) &\leqslant c' \exp\big(m_1 \log(1 + |\xi|) - \psi(\xi)\big) \sup_{|w-\xi|\leqslant 1} |G(w)| \\ &\leqslant c' \exp(m_1 + \delta) \sup_{|w-\xi|\leqslant 1} (1 + |w|)^{m_1} \exp(-\psi(w)) |G(w)| \end{aligned}$$

for every $\xi \in \mathbb{C}^n$. In particular we derive

$$(*) \qquad \begin{aligned} &\sup_{\xi \in V} |\mathcal{L}(\xi, D_\xi) G(\xi)| \exp(-\psi(\xi)) \leqslant \\ &\qquad \leqslant c' \exp(m_1 + \gamma) \sup_{\substack{|w-\xi|\leqslant 1 \\ \xi \in V}} \{(1 + |w|)^{m_1} \exp(-\psi(w)) |G(w)|\} . \end{aligned}$$

We consider now a continuous functional μ on $\Gamma(\mathbb{C}^n, \mathcal{O}_{A_0})$. We denote by η a continuous extension of μ to $\Gamma(\mathbb{C}^n, \mathcal{E}^{p_0})$ and we denote by $\tilde{\eta} = \tilde{\eta}(\xi)$ the Laplace transform of η.

Let K be a convex compact subset of $\mathbb{C}^n$ such that we have an estimate

$$(\alpha) \qquad |\mu(u)| \leqslant c(\mu) \|u\|_{K,m} \qquad \forall u \in \Gamma(\mathbb{C}^n, \mathcal{O}_{A_0}) .$$

We choose convex compact sets F_1 and F_2 in $\mathbb{C}^n$ such that

$$K \subset \mathring{F}_1 , \qquad F_1 \subset \mathring{F}_2 .$$

From Ehrenpreis fundamental principle from the estimate (α) we derive and estimate of the form

$$|\mathcal{L}(\xi, D_\xi)\tilde{\eta}(\xi)| \leqslant c_1 c(\mu)(1+|\xi|)^{m+N_1} \exp\big(H_K(\xi)\big) \qquad \forall \xi \in V$$

with $c_1 = c_1(K, m, A_0)$ and $N_1 = N_1(A_0)$. Now for some constant $c_2 = c_2(K, m, A_0)$ we have $(1+|\xi|)^{m+N_1} \exp\big(H_K(\xi)\big) \leqslant c_2 \exp\big(H_{F_1}(\xi)\big)$ for every $\xi \in \mathbb{C}^n$. Therefore we deduce an estimate

$$(\beta) \qquad |\mathcal{L}(\xi, D_\xi)\tilde{\eta}(\xi)| \leqslant c'(\mu) \exp\big(H_{F_1}(\xi)\big) \qquad \forall \xi \in V$$

and where $c'(\mu) = c_1 c_2 c(\mu)$.

On its turn from Ehrenpreis fundamental principle and the estimate (β) we deduce an estimate of the form

$$|\mu(u)| \leqslant c_3 c'(\mu) \|u\|_{F_1, N_2} \qquad \forall u \in \Gamma(\mathbb{C}^n, \mathcal{O}_{A_0})$$

with $c_3 = c_3(K, F_1)$ and $N_2 = N_2(A_0)$. Since $F_1 \subset \mathring{F}_2$ for some constant $c_4 > 0$ we have for any $f \in \Gamma(\mathbb{C}^n, \mathcal{O}^{p_0})$

$$\|f\|_{F_1, N_2} \leqslant C_4 \|f\|_{F_2} \,.$$

Therefore with $c' = c_3 c_4$ we get for any $u \in \Gamma(\mathbb{C}^n, \mathcal{O}_{A_0})$

$$\begin{aligned} |\mu(u)| &\leqslant c' c'(\mu) \|u\|_{F_2} \\ &\leqslant c' \|u\|_{F_2} \sup_{\xi \in V} |\mathcal{L}(\xi, D_\xi)\tilde{\eta}(\xi)| \exp\big(-H_{F_1}(\xi)\big) \,. \end{aligned}$$

We now apply the estimate ($*$) and we obtain with another constant $c'' = c''(K, F_1, F_2, A_0) > 0$

$$(\gamma) \qquad |\mu(u)| \leqslant c'' \|u\|_{F_2} \sup_{\xi \in V} \sup_{|w-\xi| \leqslant 1} \big\{|\tilde{\eta}(w)| \exp\big(-H_{F_1}(w)\big)\big\} \,.$$

c) Let ω be an open and convex set in $\mathbb{R}^n$ and let us assume that $H^1(\omega, \mathcal{A}_{A_0}) = 0$. By theorem 2 this is equivalent for ω to satisfy the carrier condition that we can take in the form (C_1) of proposition 5:

For every convex compact set $K \subset \omega$ we can find $\delta > 0$ and a convex compact set K' with $K \subset K' \subset \omega$ with the following property: given any convex compact set $K'' \subset \omega$, given T with $0 < T < \delta$ there exist $\varepsilon = \varepsilon(K'', T) > 0$ and $c = c(K'', T) > 0$ such that for any continuous linear functional μ on

$\Gamma(\mathbf{C}^n, \mathcal{O}_{A_0})$ *which satisfies the conditions*

$$|\mu(u)| \leqslant \|u\|_{K_{\delta+\varepsilon}} \qquad \forall u \in \Gamma(\mathbf{C}^n, \mathcal{O}_{A_0})$$

$$|\mu(u)| \leqslant \|u\|_{K''_\varepsilon} \qquad \forall u \in \Gamma(\mathbf{C}^n, \mathcal{O}_{A_0})$$

we also have that

$$|\mu(u)| \leqslant c\|u\|_{K'_T} \qquad \forall u \in \Gamma(\mathbf{C}^n, \mathcal{O}_{A_0}) .$$

Let now $F(\xi)$ be an entire function on $\mathbf{C}^n$ which satisfies the conditions (1), (2), (3) and (4) of point a) above. Then $\log|F(\xi)| - aN$ satisfies the classical Phragmén-Lindelöf principle for $\mathbf{C}^n$ (cf. proof of proposition 23 point δ)). Therefore we have

$$|F(\xi)| \leqslant \exp\big(aN + N(H_K(\xi) + \delta|\operatorname{Re}\xi|)\big) \qquad \forall \xi \in \mathbf{C}^n .$$

We introduce the function

$$G_\nu(\xi) = F^\nu\left(\frac{\xi}{\nu N}\right)$$

for any integer $\nu \geqslant 1$. Then G_ν satisfies the estimate:

(i) $|G_\nu(\xi)| \leqslant \exp\big(aN\nu + H_K(\xi) + \delta|\operatorname{Re}\xi|\big)$.

Also from condition (4) for F we deduce that for fixed $N \geqslant 1$ and $\nu \geqslant 1$, the function

(ii) $(1 + |\xi|)^k |G_\nu(\xi)| \exp\big(H_K(\xi) + \delta\,|\operatorname{Re}\xi|\big)$

is a bounded function in $\mathbf{C}^n$ for every choice of the integer $k \geqslant 0$. By Paley-Wiener theorem conditions (i) and (ii) imply that $G_\nu(\xi)$ is the Laplace transform of a C^∞ function μ_ν with support in $K_\delta \cap \mathbb{R}^n$; with usual notations we have thus

$$G_\nu(\xi) = \tilde{\mu}_\nu(\xi) .$$

d) Now let $K \subset \omega$ be given. We determine $\delta > 0$ and K' according to the carrier condition (C_1) quoted above (and which is holding because of the assumption $H^1(\omega, \mathcal{A}_{A_0}) = 0$). There is no loss of generality if we take δ so small that $K_\delta \cap \mathbb{R}^n \subset \omega$. Also enlarging if necessary K' we may assume that $K \subset \mathring{K}' \subset K' \subset \omega$. Finally for the choice of K'' it is not restrictive if we consider only those compact convex sets K'' with the property $K_\delta \cap \mathbb{R}^n \subset K''$.

With these choices of K, δ, K', K'' we have to show that any entire function which satisfies the first four conditions specified in a) also satisfies the fifth condition.

Given such entire function $F(\xi)$ we construct for every $\nu \geqslant 1$ the C^∞ function μ_ν as in point c).

Given $Y \in \mathbb{C}^{p_0}$ with $|Y| = 1$ we consider the continuous linear functional λ on $\Gamma(\mathbb{C}^n, \mathcal{O}_{A_0})$ defined by

$$\lambda(f) = \int_{\mathbb{R}^n} ({}^tYf)\mu_\nu\, dx \qquad \forall f \in \Gamma(\mathbb{C}^n, \mathcal{O}_{A_0}) ,$$

and note that we have an estimate

$$|\lambda(f)| \leqslant c(\lambda)\|f\|_{K_\delta \cap \mathbb{R}^n} , \qquad \forall f \in \Gamma(\mathbb{C}^n, \mathcal{O}_{A_0}) .$$

We can apply the arguments given in point b) above to this functional.

We first take $F_1 = K_{\delta+\varepsilon/2}$ and $F_2 = K_{\delta+\varepsilon}$ and we get from inequality (γ)

$$|\lambda(f)| \leqslant c_0 c'(\nu)\|f\|_{K_{\delta+\varepsilon}} , \qquad \forall f \in \Gamma(\mathbb{C}^n, \mathcal{O}_{A_0}) ,$$

where $c_0 = c_0(K, \delta, \varepsilon, A_0) > 0$ and where

$$c'(\nu) = \sup_{\xi \in V} \sup_{|w-\xi| \leqslant 1} |{}^tY\tilde{\mu}_\nu(w)| \exp\left(-H_{K_{\delta+\varepsilon/2}}(w)\right) .$$

Secondly we take $F_1 = K''_{\varepsilon/2}$ and $F_2 = K''_\varepsilon$ and we obtain

$$|\lambda(f)| \leqslant c_1 c''(\nu)\|f\|_{K''_\varepsilon} \qquad \forall f \in \Gamma(\mathbb{C}^n, \mathcal{O}_{A_0}) ,$$

where $c_1 = c_1(K, K'', \varepsilon, A_0) > 0$ and where

$$c''(\nu) = \sup_{\xi \in V} \sup_{|w-\xi| \leqslant 1} |{}^tY\tilde{\mu}_\nu(w)| \exp\left(-H_{K''_{\varepsilon/2}}(w)\right) .$$

From the carrier condition we derive then an inequality of the form, for any T with $0 < T < \delta$

$$|\lambda(f)| \leqslant c(T) \sup\left(c_0 c'(\nu), c_1 c''(\nu)\right)\|f\|_{K'_T} , \qquad \forall f \in \Gamma(\mathbb{C}^n, \mathcal{O}_{A_0}) .$$

e) Let now $\Theta \in V$ be fixed. We then can find $X \in \mathbb{C}^{p_0} - \{0\}$ such that $A_0(\Theta)X = 0$; this because for $\Theta \in V$ the rank of the matrix $A_0(\Theta)$ is strictly less than p_0 (cf. [4] proposition 2).

If we set

$$f = X \exp(\langle \Theta, \xi \rangle)$$

then $f \in \Gamma(\mathbb{C}^n, \mathcal{O}_{A_0})$. We can assume $|X| = 1$ without loss of generality. We choose in the considerations of the previous point d) $Y = \bar{X}$ so that we obtain

$$\lambda(f) = \int_{\mathbb{R}^n} {}^t\bar{X} X \exp(\langle \Theta, x \rangle) \mu_\nu(x)\, dx = G_\nu(\Theta)\,.$$

Using the last inequality of point d) we obtain then

$$|G_\nu(\Theta)| \leqslant c(T) \sup\big(c_0 c'(\nu), c_1 c''(\nu)\big) \exp\big(H_{K'}(\Theta) + T|\Theta|\big) \qquad \forall \Theta \in V\,,$$

since for our particular choice of f we have

$$\|f\|_{F'_T} = \sup_{\xi \in K'_T} |\exp(\langle \Theta, \xi \rangle)| = \exp\big(H_{K'}(\Theta) + T|\Theta|\big)\,.$$

In conclusion, taking $c = \sup(c_0, c_1)$ we have established the inequality

$$|G_\nu(\xi)| \leqslant c c(T) \sup\big(c'(\nu), c''(\nu)\big) \exp\big(H_{K'}(\xi) + T|\xi|\big) \qquad \forall \xi \in V\,.$$

f) From property (1) of F we derive in particular that, $\forall \xi \in \mathbb{C}^n$,

$$|G_\nu(\xi)| \leqslant \exp\big(H_K(\xi) + \delta|\xi|\big) \leqslant \exp\big(H_K(\xi) + (\delta + \varepsilon/2)|\xi|\big)\,.$$

We note also that in the definition of the constant $c'(\nu)$ we have

$$|{}^tY \tilde{\mu}_\nu(w)| = |G_\nu(w)| \qquad (\text{since } |Y| = 1)\,.$$

Therefore we deduce that

$$c'(\nu) \leqslant 1 \qquad \forall \nu \geqslant 1\,.$$

We want to evaluate the constant $c''(\nu)$. By its definition we have

$$\begin{aligned} c''(\nu) &= \sup_{\substack{w \in \mathbb{C}^n \\ \operatorname{dist}(w, V) \leqslant 1}} |G_\nu(w)| \exp\big(-H_{K''_{\varepsilon/2}}(w)\big) \\ &= \sup_{\substack{w \in \mathbb{C}^n \\ \operatorname{dist}(w, V) \leqslant 1}} \left| F\left(\frac{w}{N\nu}\right)^{\nu} \right| \exp\big(-H_{K''_{\varepsilon/2}}(w)\big) \\ &= \sup_{z \in (1/N\nu)\{w \in \mathbb{C}^n | \operatorname{dist}(w, V) \leqslant 1\}} |F(z)^\nu| \exp\big(-H_{K''_{\varepsilon/2}}(N\nu z)\big)\,. \end{aligned}$$

Therefore

$$(c''(\nu))^{1/\nu} = \sup_{z\in(1/N\nu)\{w\in\mathbf{C}^n|\operatorname{dist}(w,V)\leqslant 1\}} |F(z)| \exp\left(-H_{K''_{\varepsilon/2}}(Nz)\right).$$

Now we remark that, by the choice of δ and K'' we have

$$H_K(\xi) + \delta|\operatorname{Re}\xi| \leqslant H_{K''}(\xi) \qquad \forall \xi \in \mathbf{C}^n$$

while

$$H_{K''_{\varepsilon/2}}(\xi) = H_{K''}(\xi) + \left(\frac{\varepsilon}{2}\right)|\xi| \qquad \forall \xi \in \mathbf{C}^n .$$

By the condition (4) for F we get in particular that for

$$|z| \to \infty \qquad |F(z)| \exp\left(-H_{K''}(Nz)\right) \to 0$$

and therefore a fortiori that

$$|F(z)| \exp\left(-H_{K''_{\varepsilon/2}}(Nz)\right) \to 0 .$$

There exists therefore a real $R > 0$ such that

$$|F(z)| \exp\left(-H_{K''}(Nz)\right) \leqslant 1 \qquad \text{for } |z| \geqslant R .$$

For every $\nu \geqslant 1$ we can select $z_\nu \in \overline{B(0, R)}$ such that

$$\operatorname{dist}(N\nu z_\nu, V) \leqslant 1$$

$$|F(z_\nu)| \exp\left(-H_{K''}(Nz_\nu)\right) =$$
$$= \max\left\{|F(z)| \exp\left(-H_{K''}(Nz)\right) \middle| \, |z| \leqslant R, \operatorname{dist}(N\nu z, V) \leqslant 1\right\}.$$

We can extract from the sequence $\{z_\nu\}$ a subsequence $\{z_{\nu_h}\}$ with $z_{\nu_h} \to z_0 \in \in \overline{B(0, R)}$ for $h \to \infty$, and such that

$$\gamma = \lim_{h\to\infty} |F(z_{\nu_h})| \exp\left(-H_{K''}(Nz_{\nu_h})\right) = \max \lim_{\nu\to\infty} |F(z_\nu)| \exp\left(-H_{K''}(Nz_\nu)\right).$$

We must have

$$\gamma = |F(z_0)| \exp\left(-H_{K''}(Nz_0)\right).$$

Also $z_0 \in W$ since for every c we can find a point $\xi_\nu \in V$ with $|N\nu z_\nu - \xi_\nu| \leqslant 1$ i.e.

$$\left| z_\nu - \frac{\xi_\nu}{N\nu} \right| \leqslant \frac{1}{N\nu}$$

so that $\xi_{\nu_h}/N\nu_h \to z_0$ and $z_0 \in W$ (for $h \to \infty$). It follows then from condition (2) that $\gamma \leqslant 1$ and in particular that we have

$$\max_{\nu \to \infty} \lim c''(\nu)^{1/\nu} \leqslant 1 .$$

g) Let now $\theta \in W$. We consider a sequence $\{\xi_\nu\} \subset V$ such that, for $\nu \to \infty$, $\xi_\nu / N\nu \to \theta$. We use in ξ_ν the final inequality of point *e*). We get then

$$|G_\nu(\xi_\nu)|^{1/\nu} = \left| F\left(\frac{\xi_\nu}{N\nu}\right)\right| \leqslant$$
$$\leqslant c^{1/\nu} c(T)^{1/\nu} \sup \left(c'(\nu), c''(\nu)\right)^{1/\nu} \exp\left(NH_{K'}\left(\frac{\xi_\nu}{N\nu}\right) + NT \left|\frac{\xi_\nu}{N\nu}\right|\right) .$$

Taking the limit for $\nu \to \infty$ we derive the inequality

$$|F(\Theta)| \leqslant \exp\left(N(H_{K'}(\Theta) + T|\Theta|)\right) .$$

This inequality must hold for every T with $0 < T < \delta$. Therefore for $T \to 0$ we get

$$|F(\Theta)| \leqslant \exp\left(NH_{K'}(\Theta)\right)$$

i.e.

$$\frac{1}{N} \log |F(\xi)| \leqslant H_{K'}(\xi) \qquad \forall \xi \in W .$$

This is what we wanted to prove.

SECTION 16

Consequences of Phragmén-Lindelöf Principle

a) We consider on $\mathbb{R}^n$ a Hilbert complex

$$\mathcal{E}^{p_0}(\omega) \xrightarrow{A_0(D)} \mathcal{E}^{p_1}(\omega) \xrightarrow{A_1(D)} \mathcal{E}^{p_2}(\omega) \to \dots \tag{1}$$

defined for all open sets $\omega \subset \mathbb{R}^n$, and the corresponding Hilbert resolution

$$0 \leftarrow M \leftarrow \mathcal{P}^{p_0} \xleftarrow{{}^tA_0(\xi)} \mathcal{P}^{p_2} \xleftarrow{{}^tA_1(\xi)} \mathcal{P}^{p_1} \leftarrow \dots \tag{2}$$

where $\mathcal{P} = \mathbb{C}[\xi_1, \dots, \xi_n]$. Let $\mathfrak{p}_1, \dots, \mathfrak{p}_k$ be the prime ideals of Ass (M).

We set

$$V_i = \{\xi \in \mathbb{C}^n | p(\xi) = 0 \ \forall p \in \mathfrak{p}_i\} \qquad 1 \leqslant i \leqslant k$$

and we denote by $W = V_i^0$ the corresponding asymptotic variety which is an affine cone in $\mathbb{C}^n$ with vertex at the origin. We decompose each asymptotic variety W into irreducible components and we denote by W_j $1 \leqslant j \leqslant l$ the irreducible cones thus obtained.

Let ω be open and convex in $\mathbb{R}^n$ and let $\mathcal{A}_{A_0}$ denote the sheaf of germs of real analytic (complex valued) functions u in $\mathbb{R}^n$ which satisfy the equation $A_0(D)u = 0$.

We have established from theorem 3 and the various equivalence of Phragmén-Lindelöf principle the equivalence of the statement

$$H^1(\omega, \mathcal{A}_{A_0}) = 0$$

with the statement (Phragmén-Lindelöf principle):

Given a compact convex set $K \subset \omega$ we can find $\delta > 0$ and a compact convex set $K' \subset \omega$ with $K \subset K'$ such that for any j with $1 \leqslant j \leqslant l$ and for any plurisubharmonic function φ on $\mathbb{C}^n$ which satisfies

$$\varphi(\xi) \leqslant H_K(\xi) + \delta|\xi| \qquad \forall \xi \in W_j$$

$$\varphi(\xi) \leqslant 0 \qquad \forall \xi \in W_j \cap i\mathbb{R}^n$$

we also have

$$\varphi(\xi) \leqslant H_{K'}(\xi) \qquad \forall \xi \in W_j .$$

One deduces immediately from this statement the following useful

PROPOSITION 24. *Let $\{\omega_m\}_{m\in\mathbb{N}}$ be an increasing sequence of open convex sets in $\mathbb{R}^n$ and let $\omega = \bigcup_{m=1}^{\infty} \omega_m$.*

If for every m we have $H^1(\omega_m, \mathcal{A}_{A_0}) = 0$ then we have also $H^1(\omega, \mathcal{A}_{A_0}) = 0$.

b) For any subset $A \subset \mathbb{R}^n$, for any choice of $x_0 \in \mathbb{R}^n$ and for any $\beta > 0$ we define

$$A(x_0, \lambda) = x_0 + \lambda(A - x_0) .$$

This is the set obtained from A by the homotetic transformation with center x_0 and ratio λ.

We also define

$$\Gamma(A, x_0) = \bigcup_{\lambda>0} A(x_0, \lambda) = x_0 + \bigcup_{\lambda>0} \lambda(A - x_0) .$$

LEMMA 16. *Let ω be open and convex in $\mathbb{R}^n$. For any $x_0 \in \mathbb{R}^n$ and for any $\lambda > 0$ we have the implication*

$$H^1(\omega, \mathcal{A}_{A_0}) = 0 \Rightarrow H^1(\omega(x_0, \lambda), \mathcal{A}_{A_0}) = 0 .$$

PROOF. Given any compact (convex) set $K \subset \mathbb{R}^n$ we have

$$H_{K(x_0,\lambda)}(\xi) = \sup_{z\in K} \operatorname{Re} \langle \xi, x_0 + \lambda(z - x_0)\rangle$$
$$= \lambda H_K(\xi) + (1 - \lambda) \operatorname{Re} \langle \xi, x_0\rangle .$$

Given a compact convex set $K \subset x_0 + \lambda(\omega - x_0)$ we deduce that

$$K_1 = x_0 + \frac{1}{\lambda}(K - x_0) \subset \omega .$$

By the assumption we can find $\delta_1 > 0$ and a compact convex $K_1' \subset \omega$ with $K_1 \subset K_1'$ such that for any j with $1 \leq j \leq l$ and for any plurisubharmonic function φ in $\mathbb{C}^n$ with

$$\begin{aligned} &\varphi(\xi) \leq H_{K_1}(\xi) + \delta_1|\xi| && \forall \xi \in W_j \\ &\varphi(\xi) \leq 0 && \forall \xi \in W_j \cap i\mathbb{R}^n \end{aligned}$$

we also have

$$\varphi(\xi) \leq H_{K_1'}(\xi) .$$

Set $K' = x_0 + \lambda(K_1' - x_0) = K_1'(x_0, \lambda)$. Then $K \subset K' \subset \omega(x_0, \lambda)$.

We set $\delta = \lambda\delta_1$. We claim that Phragmén-Lindelöf principle holds in $\omega(x_0, \lambda)$ for K, K' and δ.

Indeed let ψ be a plurisubharmonic function in $\mathbb{C}^n$ such that

$$\begin{aligned} &\psi(\xi) \leq H_K(\xi) + \delta|\xi| && \forall \xi \in W_j \\ &\psi(\xi) \leq 0 && \forall \xi \in W_j \cap i\mathbb{R}^n . \end{aligned}$$

Set

$$\varphi(\xi) = \psi\left(\frac{\xi}{\lambda}\right) + \left(1 - \frac{1}{\lambda}\right) \operatorname{Re} \langle \xi, x_0 \rangle .$$

Then one has

$$\begin{aligned} \varphi(\xi) &\leq \frac{1}{\lambda} H_K(\xi) + \left(1 - \frac{1}{\lambda}\right) \operatorname{Re} \langle \xi, x_0 \rangle + \delta_1|\xi| \\ &\leq H_{K_1}(\xi) + \delta_1|\xi| \qquad \forall \xi \in W_j . \end{aligned}$$

Also one has $\varphi(\xi) = \psi(\xi/\lambda)$ for $\xi \in i\mathbb{R}^n$ so that also

$$\varphi(\xi) \leq 0 \qquad \forall \xi \in W_j \cap i\mathbb{R}^n .$$

We deduce then that

$$\psi\left(\frac{\xi}{\lambda}\right) + \left(1 - \frac{1}{\lambda}\right) \operatorname{Re} \langle \xi, x_0 \rangle \leq H_{K_1'}(\xi) \qquad \forall \xi \in W_j$$

i.e.

$$\begin{aligned} \psi(\xi) &\leq \lambda H_{K_1'}(\xi) + (1 - \lambda) \operatorname{Re} \langle \xi, x_0 \rangle \\ &\leq H_{K'}(\xi) . \end{aligned}$$

This is what we wanted to prove.

Let $\omega \subset \mathbb{R}^n$ be open and convex and let $x_0 \in \partial\omega = \bar{\omega} - \omega$. Then every half line with origin in x_0 meets ω in an open segment with one of the extremities in x_0. It follows then that for $0 < \lambda' < \lambda''$ we must have

$$\omega(x_0, \lambda') \subset \omega(x_0, \lambda'')$$

and therefore

$$(*) \qquad \Gamma(x_0, \omega) = \bigcup_{n=1}^{\infty} \omega(x_0, n)$$

is a convex cone with vertex in x_0, *the tangent cone to ω at x_0*. It may be that $\Gamma(x_0, \omega)$ is a half space: in this case $\Gamma(x_0, \omega)$ will be the tangent half space to ω at x_0 and the hyperplane which is the boundary of this half space will be the tangent hyperplane to ω at x_0.

If $\Gamma(x_0, \omega)$ is not a half space we can take $x_1 \in \partial\Gamma(x_0, \omega)$ with $x_1 \neq x_0$. Then $\Gamma(x_1, \Gamma(x_0, \omega))$ is a convex cone with vertex the line x_0x_1. If it is a half space we will call it a tangent half space to ω at x_0 and the boundary of it will be called a tangent hyperplane to ω at x_0.

If $\Gamma(x_1, \Gamma(x_0, \omega))$ is not a half space we can take $x_2 \in \partial\Gamma(x_1, \Gamma(x_0, \omega))$ with $x_2 \notin \text{line}\, x_0x_1$. Then $\Gamma\big(x_2, \Gamma(x_1, \Gamma(x_0, \omega))\big)$ will be a convex cone with vertex the plane $x_0x_1x_2$. Etc.

In this way for every point $x_0 \in \partial\omega$ we define a set of tangent half spaces to ω at x_0.

One verifies that

$$\omega = \cap\,(\text{tangent half spaces})\,.$$

Indeed if $a \notin \omega$ and $b \in \omega$ the segment ab meets ω in a segment cb which is open in c and closed in b. Thus $c \in \partial\omega$ and any tangent half space Σ to ω at c will have the property

$$a \notin \Sigma\,, \qquad \Sigma \supset \omega\,.$$

Using the remark $(*)$ made above, the previous lemma and proposition 24 we derive the following

COROLLARY 1. *Let ω be open and convex in $\mathbb{R}^n$. Let $x_0 \in \partial\omega$ and let $\Sigma(x_0)$ be a tangent half space to ω at x_0. We have the implication*

$$H^1(\omega, \mathcal{A}_{A_0}) = 0 \Rightarrow H^1(\Sigma(x_0), \mathcal{A}_{A_0}) = 0\,.$$

We also remark that for $x_0 \in \omega$ we have $\Gamma(x_0, \omega) = \mathbb{R}^n$. Thus we have also the following

COROLLARY 2. *Let ω be open and convex in $\mathbb{R}^n$, $\omega \neq \phi$. If $H^1(\omega, \mathcal{A}_{A_0}) = 0$ then also $H^1(\mathbb{R}^n, \mathcal{A}_{A_0}) = 0$.*

c) We will say that the operator $A_0(D)$ is a *terminal operator* if the Hilbert complex (2) reduces to the exact sequence

$$0 \leftarrow M \leftarrow \mathcal{P}^{p_0} \xleftarrow{{}^tA_0(\xi)} \mathcal{P}^{p_1} \leftarrow 0 .$$

This is the case for instance if $p_0 = p_1 = 1$ and $A_0(\xi)$ reduces to a non zero polynomial (a single operator in a single unknown function).

LEMMA 17. *Let $A_0(D)$ be a terminal operator. Let ω_1, ω_2 be open subsets of $\mathbb{R}^n$. We have*

$$\dim_{\mathbb{C}} H^1(\omega_1 \cap \omega_2, \mathcal{A}_{A_0}) \leqslant \dim_{\mathbb{C}} H^1(\omega_1, \mathcal{A}_{A_0}) + \dim_{\mathbb{C}} H^1(\omega_2, \mathcal{A}_{A_0}) .$$

PROOF. Denoting by $\mathcal{A}$ the sheaf of germs of real analytic complex valued functions we have the exact sequence of sheaves ([4] proposition 1)

$$0 \to \mathcal{A}_{A_0} \to \mathcal{A}^{p_0} \xrightarrow{A_0(D)} \mathcal{A}^{p_1} \to 0 .$$

Therefore for any open set $\omega \subset \mathbb{R}^n$ we have $H^2(\omega, \mathcal{A}_{A_0}) = 0$.

From the Mayer-Vietoris sequence we deduce the exact sequence

$$H^1(\omega_1, \mathcal{A}_{A_0}) \oplus H^1(\omega_2, \mathcal{A}_{A_0}) \to H^1(\omega_1 \cap \omega_2, \mathcal{A}_{A_0}) \to H^2(\omega_1 \cup \omega_2, \mathcal{A}_{A_0}) .$$

Since $H^2(\omega_1 \cup \omega_2, \mathcal{A}_{A_0}) = 0$ we deduce the statement of the lemma.

For terminal operators the implication of corollary 1 above can be inverted at least for convex open bounded sets ω:

PROPOSITION 25. *Let $A_0(D)$ be a terminal operator. Let $\omega \subset \mathbb{R}^n$ be open bounded and convex. Assume that for any tangent half space Σ to ω we have $H^1(\Sigma, \mathcal{A}_{A_0}) = 0$. Then one has also $H^1(\omega, \mathcal{A}_{A_0}) = 0$.*

PROOF. α) Without loss of generality we may assume that the origin of the coordinates in $\mathbb{R}^n$ is contained in ω.

Let K be a compact subset of ω. We can find $\varepsilon > 0$ so small that

$$K \subset \frac{1}{1+\varepsilon}\,\omega .$$

We claim that we can find a polyhedron P with the following properties

i) P is the intersection of finite many tangent half spaces to ω;

ii) P is contained and relatively compact in $(1+\varepsilon)\omega$.

Indeed set $\omega_1 = (1+\varepsilon)\omega$. Let $x_0 \in \partial\omega_1$. We can find a linear polynomial $l_{x_0}(x) = \sum a_i x_i + b$ with the following properties

(a) $l_{x_0}(x) < 0$ is a tangent half space to ω

(b) $l_{x_0}(x_0) > 0$.

Let $U(x_0) = \{x \in \partial\omega_1 | l_{x_0}(x) > 0\}$. This is an open neighborhood of x_0 in $\partial\omega_1$. By the compacity of $\partial\omega_1$ we can find a finite set $x_0^{(1)}, x_0^{(2)}, \dots, x_0^{(k)}$ of points in $\partial\omega_1$ such that

$$\partial\omega_1 \subset \bigcup_{j=1}^{k} U(x_0^{(j)}) .$$

We define then

$$P = \{x \in \mathbb{R}^n | l_{x_0^{(1)}}(x) < 0, \dots, l_{x_0^{(k)}}(x) < 0\} .$$

Then P has the required properties.

β) Because of lemma 17 (with repeated applications of it) we have

$$H^1(P, \mathcal{A}_{A_0}) = 0 .$$

Because of lemma 16 we also have

$$H^1\left(\frac{1}{1+\varepsilon} P, \mathcal{A}_{A_0}\right) = 0 .$$

Now $\big(1/(1+\varepsilon)\big)\omega \subset \big(1/(1+\varepsilon)\big)P \subset \omega$.

Because of Phragmén-Lindelöf principle for $\big(1/(1+\varepsilon)\big)P$ we deduce that we can find $\delta > 0$ and a compact set $K' \subset \big(1/(1+\varepsilon)\big)P$ with $K \subset K'$ such that, for any j, $1 \leqslant j \leqslant l$, and for any plurisubharmonic function φ on $\mathbb{C}^n$ verifying

$$\varphi(\xi) \leqslant H_K(\xi) + \delta|\xi| \qquad \forall \xi \in W_j$$

$$\varphi(\xi) \leqslant 0 \qquad \forall \xi \in W_j \cap i\mathbb{R}^n$$

we also have

$$\varphi(\xi) \leqslant H_{K'}(\xi) \qquad \forall \xi \in W_j .$$

This shows that Phragmén-Lindelöf principle holds for ω and therefore we must have $H^1(\omega, \mathcal{A}_{A_0}) = 0$.

d) From corollary 1 given above follows that the first investigation on analytic convexity should be directed towards the simple case $\omega = \mathbb{R}^n$. We note that for $\mathbb{R}^n$ Phragmén-Lindelöf principle can be formulated as follows: « There is a constant $c > 0$ such that, for any j with $1 \leqslant j \leqslant l$, and any plurisubharmonic function such that

$$\varphi(\xi) \leqslant |\xi| \qquad \forall \xi \in W_j$$

$$\varphi(\xi) \leqslant 0 \qquad \forall \xi \in W_j \cap i\mathbb{R}^n$$

we also have

$$\varphi(\xi) \leqslant c|\mathrm{Re}\, \xi| \qquad \forall \xi \in W_j \text{ »}.$$

Indeed this is the principle with $K = \{0\}$, $\delta = 1$, $K' = \{x \in \mathbb{R}^n \big| \, |x| \leqslant c\}$.

Conversely if the above statement holds and if φ is a plurisubharmonic function which satisfies for some $\delta > 0$ and some compact K

$$\varphi(\xi) \leqslant H_K(\xi) + \delta|\xi| \qquad \forall \xi \in W_j$$

$$\varphi(\xi) \leqslant 0 \qquad \forall \xi \in W_j \cap i\mathbb{R}^n$$

we derive, if $K \subset \{x \in \mathbb{R}^n \big| \, |x| \leqslant R\}$, that $\varphi(\xi) \leqslant (R + \delta)|\xi|$.

Thus $(R + \delta)^{-1}\varphi$ satisfies the above statement and we thus conclude with $K' = \{x \in \mathbb{R}^n \big| \, |x| \leqslant (R + \delta)c\}$, that

$$\varphi(\xi) \leqslant H_{K'}(\xi) \qquad \forall \xi \in W_j .$$

SECTION 17

Algebraic Cones with Phragmén-Lindelöf Principle

a) Let W be an algebraic cone in $\mathbb{C}^n$ (with vertex at the origin). *We will say that W admits Phragmén-Lindelöf principle* if we can find a constant $c>0$ such that:

any plurisubharmonic function $\varphi(\xi)$ defined in $\mathbb{C}^n$ and verifying

$$\varphi(\xi) \leqslant |\xi| \qquad \forall \xi \in W$$

$$\varphi(\xi) \leqslant 0 \qquad \forall \xi \in W \cap i\mathbb{R}^n$$

verifies also the inequality

$$\varphi(\xi) \leqslant c|\mathrm{Re}\,\xi| \qquad \forall \xi \in W\,.$$

If a Hilbert complex (1) of differential operators with constant coefficients is given (as in the previous section) and if for some non empty open convex set ω we have $H^1(\omega, \mathcal{A}_{A_0}) = 0$ then each asymptotic variety W_j $1 \leqslant j \leqslant l$ associated (as explained in the previous section) to the given complex will admit the Phragmén-Lindelöf principle.

Indeed because of corollary 2 to lemma 16 we will have

$$H^1(\mathbb{R}^n, \mathcal{A}_{A_0}) = 0\,.$$

This fact entails the condition given above for any cone W_j $1 \leqslant j \leqslant l$ as explained in the previous section point *d*).

b) Given the algebraic cone W we consider its real part

$$W_{\mathbb{R}} = W \cap \mathbb{R}^n\,.$$

For any $\lambda \in \mathbb{C}$ we have $\lambda W \subset W$ and if $\lambda \neq 0$ $\lambda W = W$ since W is a cone.

We have therefore

$$W \cap i\mathbb{R}^n = i(W \cap \mathbb{R}^n) = iW_{\mathbb{R}}$$

therefore the real part $W_{\mathbb{R}}$ of W is carried by the multiplication by i (which is an isomorphism of W onto itself) into the region $W \cap i\mathbb{R}^n$ that occur in the enunciation of Phragmén-Lindelöf principle.

PROPOSITION 26. *Let W be an elgebraic cone in $\mathbb{C}^n$ which admits the Phragmén-Lindelöf principle.*

Let $a \in W_{\mathbb{R}}$ with $a \neq 0$ and let (W_α, a) denote any irreducible germ of W at the point a. Then we have

$$\dim_{\mathbb{R}} (W_\alpha \cap \mathbb{R}^n, a) = \dim_{\mathbb{C}} (W_\alpha, a) .$$

PROOF. α) Let $(W, a) = (W_\alpha, a) \cup (W_\beta, a) \cup \ldots \cup (W_\gamma, a)$ be the decomposition of the germ (W, a) into irreducible components. Assume that for some irreducible germ, say (W_α, a), we have

$$\dim_{\mathbb{R}} (W_\alpha \cap \mathbb{R}^n, a) < \dim_{\mathbb{C}} (W_\alpha, a) .$$

Consider the real analytic germ $(W_\alpha \cap \mathbb{R}^n, a)$ and denote by (Σ, a) its complexification (cf. [16] p. 91-93). We have

$$\dim_{\mathbb{C}} (\Sigma, a) = \dim_{\mathbb{R}} (W_\alpha \cap \mathbb{R}^n, a) .$$

Also we must have $(\Sigma, a) \subset (W_\alpha, a)$, by the properties of the complexification. Since the dimensions are different we have $(\Sigma, a) \subsetneqq (W_\alpha, a)$.

We conclude then that if the equality of the proposition is violated we can find a proper analytic subgerm $(Z, a) \subsetneqq (W, a)$ with $(W \cap \mathbb{R}^n, a) \subset (Z, a)$. Therefore we can find a germ of a holomorphic function at a which vanishes on $(W \cap \mathbb{R}^n, a)$ but does not vanish on (W, a).

To prove the proposition is therefore enough to prove the following statement:

any germ of a holomorphic function on W at a, which vanishes on the germ $(W \cap \mathbb{R}^n, a)$, vanishes automatically on the whole germ (W, a).

β) Set $b = ia$. We may assume as well $|b| = |\mathrm{Im}\, b| = 1$. We transform the above statement by the automorphism of W which is the multiplication by i. We have to show that any germ of holomorphic function w on W at b, which vanishes on the germ $(W \cap i\mathbb{R}^n, b)$, vanishes automatically on the whole germ (W, b).

Let r_0 with $0 < r_0 < 1$ be so chosen that w is defined and holomorphic on the set $W \cap \overline{B(b, r_0)}$. We also select r with $0 < r < r_0$.

Assume that W satisfies the Phragmén-Lindelöf principle with the constant c.

According to proposition 16 we can find a constant $c' > 0$ such that for any choice of $\Theta \in W$ with $|\mathrm{Im}\,\Theta| = 1$ $|\mathrm{Re}\,\Theta| < r/2$ and for any choice of a plurisubharmonic function φ on $W \cap B(\Theta, r)$ which satisfies

$$\varphi(\xi) \leqslant |\xi| \qquad \forall \xi \in W \cap B(\Theta, r)$$

$$\varphi(\xi) \leqslant 0 \qquad \forall \xi \in W \cap i\mathbb{R}^n \cap B(\Theta, r)$$

we have also that

$$\varphi(\Theta) \leqslant c' |\mathrm{Re}\,\Theta| \qquad \forall \xi \in W \cap B(\Theta, r)\,.$$

Let $\varepsilon > 0$ with $c'(\varepsilon/(1-\varepsilon)) < (1-r)/4$. We fix $\xi \in W \cap B(b, \varepsilon)$. Since $b \in i\mathbb{R}^n$ and $|b| = 1$, we deduce that

$$|\mathrm{Re}\,\xi| < \varepsilon \qquad 1-\varepsilon < |\mathrm{Im}\,\xi| < 1 + \varepsilon\,.$$

We have $\mathrm{Im}\,\xi \neq 0$ (if $\varepsilon < 1$). We define

$$\Theta = \frac{\xi}{|\mathrm{Im}\,\xi|}$$

so that

$$|\mathrm{Re}\,\Theta| < \frac{\varepsilon}{1-\varepsilon} < \frac{r}{2} \qquad \text{if } \varepsilon \text{ is small}$$

and

$$|\mathrm{Im}\,\Theta| = 1\,.$$

We now consider on $B(\Theta, r)$ the function (for any $\sigma > 0$)

$$\varphi(\eta) = \frac{1-r}{2} + \sigma \log \big|w(|\mathrm{Im}\,\xi|\eta)\big|\,.$$

This is well defined because, for ε sufficiently small, we have

$$\begin{aligned} \big||\mathrm{Im}\,\xi|\eta - b\big| &\leqslant \big||\mathrm{Im}\,\xi|\eta - |\mathrm{Im}\,\xi|\Theta\big| + |\xi - b| \\ &\leqslant (1+\varepsilon)r + \varepsilon < r_0 \end{aligned}$$

if $\eta \in B(\Theta, r)$.

We note that for $\eta \in B(\Theta, r)$ we have

$$|\eta| > 1 - r$$

since $|\Theta| \geqslant |\text{Im}\,\Theta| = 1$.

Now $|w|$ is bounded on $B(b, r_0) \cap W$. Therefore for σ sufficiently small we have

$$\varphi(\eta) \leqslant 1 - r \leqslant |\eta| \qquad \eta \in W \cap B(\Theta, r)\,.$$

Also we do have

$$\varphi(\eta) \leqslant 0 \qquad \text{for } \eta \in W \cap i\mathbb{R}^n \cap B(\Theta, r)$$

because if $\eta \in W \cap i\mathbb{R}^n$ also $|\text{Im}\,\xi|\eta \in W \cap i\mathbb{R}^n$ and there w vanishes so that, for those η, $\varphi(\eta) = -\infty$.

Now $\varphi(\eta)$ is plurisubharmonic on $B(\Theta, r)$. We deduce then that (according to the above form of Phragmén-Lindelöf principle)

$$\varphi(\Theta) \leqslant c'|\text{Re}\,\Theta|$$

i.e.

$$\frac{1-r}{2} + \sigma \log |w(\xi)| \leqslant c' \frac{\varepsilon}{1-\varepsilon} < \frac{1-r}{4}\,.$$

Now this conclusion should be valid for any $\sigma > 0$ sufficiently small.

It follows that we must have for any $\xi \in W \cap B(b, \varepsilon)$

$$w(\xi) = 0\,,$$

otherwise taking the limit for $\sigma \to 0$ we deduce a contradiction. We have thus proved that w vanishes in a sufficiently small neighborhood of b. This proves our contention.

b) Let W be an algebraic cone in $\mathbb{C}^n$ which admits the Phragmén-Lindelöf principle and let $a \in W_{\mathbb{R}}$, $a \neq 0$. Consider the antinvolution j on $\mathbb{C}^n$ defined by the conjugation of the coordinates:

$$j(\xi) = \bar{\xi}\,.$$

Let (W_α, a) be any irreducible germ of W at a. We claim that *the antinvolution j sends (W_α, a) into itself.*

Indeed, let us consider the real algebraic germ $(W_\alpha \cap \mathbb{R}^n, a)$. This is left pointwise fixed by j. Let (B_α, a) denote the complexification of this last germ. By its very definition, (B_α, a) is changed into itself by j. Now by construction we must have $(B_\alpha, a) \subset (W_\alpha, a)$. But (W_α, a) is irreducible and, because of proposition 26

$$\dim_{\mathbb{C}} (B_\alpha, a) = \dim_{\mathbb{R}} (W_\alpha \cap \mathbb{R}^n, a) = \dim_{\mathbb{C}} (W_\alpha, a)\,,$$

therefore we must have $(B_\alpha, a) = (W_\alpha, a)$ and our assertion is proved. Therefore at each point $a \in W_{\mathbb{R}}$ with $a \neq 0$ the antinvolution j sends each irreducible germ (W_α, a) of W at a into itself and on each irreducible germ (W_α, a) the set of fixed points $(W_{\alpha\mathbb{R}}, a)$ is a real analytic set with

$$\dim_{\mathbb{R}} (W_{\alpha\mathbb{R}}, a) = \dim_{\mathbb{C}} (W_\alpha, a)\,.$$

Let $a \in W_{\mathbb{R}} - \{0\}$ and let

$$(W, a) = (W_1, a) \cup \ldots \cup (W_l, a)$$

be the decomposition of the germ (W, a) into irreducible components. For each s with $1 \leqslant s \leqslant l$ let

$$(W_s^*, b_s) \xrightarrow{\pi} (W_s, a)$$

be the normalization of (W_s, a).

Let $\tau = j|W_s$; then (by Riemann extension theorem) there exists a unique antinvolution σ of (W_s^*, b_s) into itself such that we have a commutative diagram

$$\begin{array}{ccc} (W_s^*, b_s) & \xrightarrow{\sigma} & (W_s^*, b_s) \\ \downarrow \pi & & \downarrow \pi \\ (W_s, a) & \xrightarrow{\tau} & (W_s, a)\,. \end{array}$$

In particular $\sigma(b_s) = b_s$.

We conclude then with the following statement.

Let

$$W = W_1 \cup \ldots \cup W_k$$

be the decomposition of W into irreducible components.

If $W_\alpha \cap \mathbb{R}^n \neq \{0\}$ *then*

i) W_α *is changed into itself by the conjugation* j;

ii) *if* $\tau = j|W_\alpha$ *and if* $W^*_\alpha \xrightarrow{\pi} W_\alpha$ *is the normalization of* W_α, *there exists a unique antinvolution* $\sigma \colon W^*_\alpha \to W^*_\alpha$ *such that* $\pi \circ \sigma = \tau \circ \pi$;

iii) *if* $W^*_{\alpha\mathbb{R}}$ *is the fixed set of* σ *and* $W_{\alpha\mathbb{R}}$ *the fixed set of* τ *we have*

$$W^*_{\alpha\mathbb{R}} = \pi^{-1}(W_{\alpha\mathbb{R}})$$

iv) *at each point* $b \in W^*_{\alpha\mathbb{R}}$ *we also have*

$$\dim_{\mathbb{R}} (W^*_{\alpha\mathbb{R}}, b) = \dim_{\mathbb{C}} W^*_\alpha .$$

c) Let now $a \in W_{\mathbb{R}} - \{0\}$ be fixed. Each irreducible component W_α of W that contains a is in the condition specified above.

We can assume that $|a| = 1$ and we can choose $r_0 > 0$ with $r_0 < 1$ such that the ball $B(a, r_0)$ meets only those irreducible components of W that contain a. Let us consider the automorphism of W given by the multiplication by i. Set

$$\alpha = ia .$$

Then the ball $B(\alpha, r_0)$ meets only irreducible components of W that contain α. The antinvolution j is changed into the antinvolution j' where

$$j'(z) = -\bar{z}$$

and the fixed set in $\mathbb{C}^n$ is $i\mathbb{R}^n$. All the arguments given in *b*) can be repeated replacing j by j'.

Consider the analytic set

$$A = W \cap B(\alpha, r_0)$$

and its normalization $A^* \xrightarrow{\pi} A$.

If $\tau = j'|A$ then τ changes A into itself. We can then find an antinvolution $\sigma \colon A^* \to A^*$ such that $\tau \circ \sigma = \tau \circ \pi$.

Now A and therefore A^* is a Stein space and we may assume without loss of generality that

i) A^* is imbedded as an analytic subset of some open set D in some numerical space $\mathbb{C}^N$;

ii) D is invariant by the antinvolution $l: w \to -\overline{w}$ where w denote the coordinates of $\mathbb{C}^N$;

iii) A^* is changed into itself by l and $l|A^* = \sigma$.

Let now $\xi \in A$: we define

$$\varphi(\xi) = \sup\{|\mathrm{Re}\, w| \,|w \in \pi^{-1}(\xi)\}\,.$$

Then $\varphi(\xi)$ is a weakly plurisubharmonic function defined on A, since $|\mathrm{Re}\, w|$ is plurisubharmonic on A^*.

We now use the following local form of Phragmén-Lindelöf principle. We choose r with $0 < r < r_0$ and let ϱ with $0 < \varrho < r$.

« There exists a constant $c > 0$ such that for any choice of $\alpha \in W \cap i\mathbb{R}^n$ with $|\alpha| = 1$ and for any weakly plurisubharmonic function φ defined on $W \cap B(\alpha, r)$ and such that

$$\varphi(\xi) \leqslant |\xi| \qquad \forall \xi \in W \cap B(\alpha, r)$$

$$\varphi(\xi) \leqslant 0 \qquad \forall \xi \in W \cap i\mathbb{R}^n \cap B(\alpha, r)$$

we also have

$$\varphi(\xi) \leqslant c|\mathrm{Re}\, \xi| \qquad \forall \xi \in W \cap B(\alpha, \varrho)\text{ »}.$$

Indeed let $r_1 = (r - \varrho)/(1 + \varrho)$. According to proposition 16 we can find a constant $c_1 > 0$ such that « if $\Theta \in W$, $|\mathrm{Im}\, \Theta| = 1$ and φ is a weakly plurisubharmonic function defined on $W \cap B(\Theta, r_1)$ and such that

$$\varphi(\xi) \leqslant |\xi| \qquad \forall \xi \in W \cap B(\Theta, r_1)$$

$$\varphi(\xi) \leqslant 0 \qquad \forall \xi \in W \cap i\mathbb{R}^n \cap B(\Theta, r_1)$$

we also have

$$\varphi(\Theta) \leqslant c|\mathrm{Re}\, \Theta|\text{ »}.$$

We take now $\xi \in W \cap B(\alpha, \varrho)$ and apply the above principle to the function

$$\psi(\eta) = |\mathrm{Im}\, \xi|^{-1} \varphi(|\mathrm{Im}\, \xi|\eta)$$

which is weakly plurisubharmonic and defined on $B(\xi/|\mathrm{Im}\, \xi|, r_1)$, taking $\Theta = \xi/|\mathrm{Im}\, \xi|$. We then conclude with the first of these statements.

Set

$$M = \sup_{w \in \pi^{-1}(W \cap B(\alpha, r))} |\mathrm{Re}\, w| \,.$$

Since $W \cap B(\alpha, r)$ is relatively compact in A and since π is proper we deduce that M is finite.

On $B(\alpha, r)$ we have

$$|\xi| \geqslant |\alpha| - r = 1 - r \,.$$

Therefore for the function φ defined above on A we have

$$\varphi(\xi) \leqslant \frac{M}{1-r} |\xi| \qquad \forall \xi \in W \cap B(\alpha, r)$$

$$\varphi(\xi) \leqslant 0 \qquad \forall \xi \in W \cap i\mathbb{R}^n \cap B(\alpha, r)$$

because if ξ is purely imaginary each point $w \in \pi^{-1}(\xi)$ is also purely imaginary.

We conclude therefore that with a positive constant c depending on r and M we must have

$$(*) \qquad \varphi(\xi) \leqslant c|\mathrm{Re}\, \xi| \qquad \forall \xi \in W \cap B(\alpha, \varrho) \,.$$

d) Let

$$\pi^{-1}(\alpha) = \beta_1 \cup \ldots \cup \beta_k \,.$$

If we take r_0 sufficiently small we may assume that the β_j's belong to two by two distinct irreducible components A_j^* of A^* so that

$$A^* = A_1^* \cup \ldots \cup A_k^* \cup L$$

and L is an analytic set disjoint from $A_1^*, \ldots, A_k^*$.

Let us assume that one of the points of $\pi^{-1}(\alpha)$, β_1 for instance, is a non singular point.

Then, if r_0 is taken sufficiently small, we can assume that A_1^* is a d-dimensional manifold. In fact it is not restrictive to assume that A_1^* is an open neighborhood of the origin in the subspace $\mathbb{C}^d$ of $\mathbb{C}^N$ spanned by the first d coordinates $w_1, \ldots, w_d$.

Let $\pi(A_1^*) = A_1$; this is an irreducible component of the set A. We claim that the germ (A_1, α) is non singular.

Let

$$\begin{aligned} \xi_1 &= g_1(w_1, \ldots, w_d) \\ &\cdots\cdots\cdots\cdots \\ \xi_n &= g_n(w_1, \ldots, w_d) \end{aligned}$$

be a set of equations for the map $\pi | A_1^*$. It will be enough to prove that

$$\operatorname{rank}\left\{\frac{\partial(g_1, \ldots, g_n)}{\partial(w_1, \ldots, w_d)}\right\}_{w=0} = d\,.$$

We remark that the g_α's are holomorphic functions defined near the origin and by the assumptions must satisfy the condition

$$g(-\overline{w}) = -\overline{g(w)}\,.$$

If $\Sigma a_\alpha w^\alpha$ is the Taylor expansion of g at the origin we must have

$$(-1)^{|\alpha|} a_\alpha = -\overline{a}_\alpha\,.$$

In particular (for $|\alpha| = 1$) we deduce that the partial derivatives of the g's at the origin must be real.

Assume, if possible, that the rank of the above jacobian matrix is strictly less than d. We can then find constants $\lambda_1, \ldots, \lambda_d$, not all zero, such that

$$\sum \lambda_j \left(\frac{\partial g_h}{\partial w_j}\right)_0 = 0 \qquad \text{for } 1 \leqslant h \leqslant n\,.$$

Because of the fact that the numbers $(\partial g_h / \partial w_j)_0$ are real, we can assume without loss of generality that the λ's are real. Let for instance $\lambda_1 \neq 0$. We then make the real change of coordinates

$$\begin{aligned} w_1 &= \lambda_1 w_1' \\ w_2 &= \lambda_2 w_1' + w_2' \\ &\cdots\cdots\cdots\cdots \\ w_d &= \lambda_d w_1' + w_d'\,. \end{aligned}$$

In the new coordinates we then have

$$\left(\frac{\partial g_h}{\partial w_1'}\right)_0 = 0 \qquad \text{for } 1 \leqslant h \leqslant n\,,$$

so that we will have expressions of the form:

$$(**) \qquad g_h(w') = ia_h + \sum_2^d b_h^j w_j' + 0(|w|^2)$$
$$1 \leqslant h \leqslant n .$$

We consider the point, for $\varepsilon > 0$ and sufficiently small,

$$w' = (\varepsilon, 0, ..., 0)$$

and let $\xi(\varepsilon) = \pi(w')$. By the definition of the function φ (with respect to this last choice of the coordinates w' on A_1^*) we will have

$$\varepsilon \leqslant \varphi(\xi(\varepsilon)) .$$

From the other hand, because of the expression $(**)$ we will have

$$|\mathrm{Re}\, \xi(\varepsilon)| = \{\Sigma(\mathrm{Re}\, g_h(\varepsilon, 0, ..., 0))^2\}^{\frac{1}{2}} = O(\varepsilon^2) .$$

But from the inequality $(*)$ at the end of point $c)$ we derive that

$$\varepsilon \leqslant cO(\varepsilon^2) .$$

This is impossible and therefore the rank of the jacobian matrix $(\partial(g)/\partial(w))_0$ is equal d. This proves our contention.

We have therefore established the following

PROPOSITION 27. *Let W be an algebraic cone in $\mathbb{C}^n$ which admits the Phragmén-Lindelöf principle.*

Let $a \in W_{\mathbb{R}}$ with $a \neq 0$, let (W_α, a) be an irreducible germ of W at a and let $(W_\alpha^*, b) \to (W_\alpha, a)$ be the normalization of it.

If the germ (W_α^*, b) is non singular at b then also the germ (W_α, a) is non singular at a.

COROLLARY 1. *Let W be as above and assume that $\dim_{\mathbb{C}} W = 2$. Then for every $a \in W_{\mathbb{R}}$ with $a \neq 0$, any irreducible germ (W_α, a) of W at a is non singular.*

PROOF. If $\dim_{\mathbb{C}} (W_\alpha, a) = 1$ there is nothing to prove as W is there a straight line. Let $\dim_{\mathbb{C}} (W_\alpha, a) = 2$.

Then in a sufficiently small neighborhood of a W_α is the product of the disc $D = \{t \in \mathbb{C}\, |t| < \varepsilon\}$ and an analytic curve Γ (as $W - \{0\}$ is a locally trivial fiber space over the curve at infinity of the cone W);

$$W_\alpha \simeq D \times \Gamma .$$

The normalization of a product is the product of the normalizations so that, if $\Gamma^* \xrightarrow{\pi} \Gamma$ is the normalization of Γ, we have

$$W^*_\alpha \simeq D \times \Gamma^*$$

(cf. [12] Satz. 14 p. 197).

But Γ^* is non singular and thus W^*_α is non singular. It follows then by the previous proposition that (W_α, a) is non singular. Let us denote by W' the union of the irreducible components of W that contain some point $a \in W_{\mathbf{R}}$ with $a \neq 0$. By the last remark at point b) above, W' is changed into itself by the conjugation j of the coordinates of $\mathbf{C}^n$ and the fixed set of $\sigma = j|W'$ is a real analytic set $W'_{\mathbf{R}}$.

At each point $a \in W'_{\mathbf{R}}$ each irreducible complex germ (W_α, a) of W' has the property

$$\dim_{\mathbf{R}}(W_{\alpha,\mathbf{R}}, a) = \dim_{\mathbf{C}}(W_\alpha, a) .$$

Let $W'^* \to W'$ be the normalization of W'. There is a unique antinvolution σ on W'^* such that

$$\pi \circ \sigma = \tau \circ \pi .$$

The fixed set of σ, $W'^*_{\mathbf{R}}$, has the property that

$$W'^*_{\mathbf{R}} = \pi^{-1}(W'_{\mathbf{R}}) .$$

Now the singular set $S(W'^*)$ of W'^* is of complex codimension $\geqslant 2$ at each one of its points. It follows that for each (irreducible) germ W^*_α of W'^* the singular set of it $S(W^*_\alpha)$ has codimension $\geqslant 2$ and therefore $S(W^*_\alpha) \cap \cap W^*_{\alpha,\mathbf{R}} = S(W^*_\alpha)_{\mathbf{R}}$ is of real codimension $\geqslant 2$ in $W^*_{\alpha,\mathbf{R}}$. Since π is proper and with discrete fibers π preserves dimension. We have therefore the following

COROLLARY 2. *Let $a \in W_{\mathbf{R}}$ with $a \neq 0$ and let W_α be any irreducible germ of W at a.*

Let $S(W_\alpha)_{\mathbf{R}}$ denote the set of points of $W_{\alpha,\mathbf{R}}$ which are singular points of W_α. Then the real codimension of $(S(W_\alpha)_{\mathbf{R}}, a)$ in $(W_{\alpha,\mathbf{R}}, a)$ is greater or equal 2.

SECTION 18

Locally Hyperbolic Cones

a) We consider in $\mathbb{C}^d$ the polycylindrical norm

$$\|w\| = \sup_{1\leqslant i\leqslant d} |w_i| \,,$$

we set, for $\sigma > 0$, $P(\sigma) = \{w \in \mathbb{C}^d \,|\, \|w\| < \sigma\}$.

LEMMA 18. *We can find a constant $c > 0$ such that for every plurisubharmonic function defined on $P(\varrho)$, $\varrho > 0$, and verifying the conditions*

$$\varphi(w) \leqslant 1 \qquad \forall w \in P(\varrho)$$

$$\varphi(w) \leqslant 0 \qquad \forall w \in P(\varrho) \cap i\mathbb{R}^n$$

we also have

$$\varphi(w) \leqslant \frac{c}{\varrho} \sum_1^d |\mathrm{Re}\, w_i| \qquad \forall w \in P\left(\frac{\varrho}{2}\right).$$

PROOF. α) It is enough to establish this lemma for $d = 1$.

In fact if this is proved, given any φ satisfying the above stated conditions, we fix $w_2, \ldots, w_d$ purely imaginary with $|w_j| < \varrho$ $2 \leqslant j \leqslant d$ and apply the lemma to the function of one variable z, for $|z| < \varrho$,

$$\psi_1(z) = \varphi(z, w_2, \ldots, w_d) \,.$$

We deduce then that

$$\psi_1(z) \leqslant \frac{c}{\varrho} |\mathrm{Re}\, z| \qquad \text{for } |z| < \frac{\varrho}{2} \,.$$

We consider then $w_1, w_3, \ldots, w_d$ fixed with

$$|w_1| < \varrho/2 \text{ and } w_3, \ldots, w_d \text{ purely imaginary with } |w_j| < \varrho,\ 3 \leqslant j \leqslant d \,.$$

The function of one variable z, for $|z| < \varrho$

$$\psi_2(z) = \varphi(w_1, z, w_3, \ldots, w_d) - \frac{c}{\varrho}|\mathrm{Re}\, w_1|$$

satisfies the hypothesis of the lemma. We deduce then that

$$\psi_2(z) \leqslant \frac{c}{\varrho}|\mathrm{Re}\, z| \quad \text{for } |z| < \frac{\varrho}{2}$$

so that

$$\varphi(w_1, w_2, w_3, \ldots, w_d) \leqslant \frac{c}{\varrho}\{|\mathrm{Re}\, w_1| + |\mathrm{Re}\, w_2|\}$$

for $|w_1| < \varrho/2$, $|w_2| < \varrho/2$ and $w_3, \ldots, w_d$ fixed purely imaginary with $|w_j| < \varrho$, $3 \leqslant j \leqslant d$.

We consider next $w_1, w_2, w_4, \ldots, w_d$ fixed with $|w_1| < \varrho/2$, $|w_2| < \varrho/2$ and $w_4, \ldots, w_d$ purely imaginary with $|w_j| \leqslant \varrho$, $4 \leqslant j \leqslant d$.

The function of one variable

$$\psi_3(z) = \varphi(w_1, w_2, z, w_4, \ldots, w_d) - \frac{c}{\varrho}\{|\mathrm{Re}\, w_1| + |\mathrm{Re}\, w_2|\}$$

satisfies the conditions of the lemma so that we must have

$$\psi_3(z) \leqslant \frac{c}{\varrho}|\mathrm{Re}\, z| \quad \text{for } |z| < \frac{\varrho}{2}.$$

Hence

$$\varphi(w_1, w_2, w_3, w_4, \ldots, w_{Sd}) \leqslant \frac{c}{\varrho}\{|\mathrm{Re}\, w_1| + |\mathrm{Re}\, w_2| + |\mathrm{Re}\, w_3|\}$$

for $|w_1| < \varrho/2$, $|w_2| < \varrho/2$, $|w_3| < \varrho/2$ and $w_4, \ldots, w_d$ purely imaginary with $|w_j| < \varrho$, $4 \leqslant j \leqslant d$.

Iterating this argument we conclude with the statement of the lemma for any number d of variables.

β) We now establish the lemma for $d = 1$. The change of variable $z = \varrho w$ reduces the proof to the case $\varrho = 1$.

Let us therefore assume that φ is subharmonic in the unit disc

$$D = \{|z| < 1\}$$

and satisfies the condition

$$\varphi(z) \leqslant 1 \quad \text{for } |z| < 1$$

$$\varphi(z) \leqslant 0 \quad \text{for } \operatorname{Re} z = 0,\ |z| < 1 .$$

Set

$$D^- = \{|z| \leqslant 1, \operatorname{Re} z \leqslant 0\}, \qquad D^+ = \{|z| \leqslant 1, \operatorname{Re} z \geqslant 0\} .$$

We consider the harmonic functions

$$\frac{2}{\pi} \arg \frac{z+i}{i-z} \qquad \text{on } D^- - \{i\} - \{-i\}$$

and

$$\frac{2}{\pi} \arg \frac{i-z}{z+i} \qquad \text{on } D^+ - \{i\} - \{-i\} .$$

The first (second) of these functions takes the value 0 on the segment $\{|z| < 1, \operatorname{Re} z = 0\}$ and the value 1 on the half circle $\{|z| = 1, \operatorname{Re} z < 0\}$ (respectively $\{|z| = 1, \operatorname{Re} z > 0\}$).

Set

$$w = \frac{z+i}{i-z} .$$

This defines a conformal map of the region $D^- - \{i\} - \{-i\}$ onto the first quadrant $Q^+ = \{\operatorname{Re} w \geqslant 0, \operatorname{Im} w \geqslant 0\} - \{0\}$.

Consider $\varphi(z)$ as a function of w on Q^+,

$$\psi(w) = \varphi\left(i \frac{w-1}{w+1}\right) .$$

We have

$$\psi(w) \leqslant 0 \quad \text{if } w \in \mathbb{R}\ w > 0$$

$$\psi(w) \leqslant 1 \quad \text{if } w \in Q^+ .$$

Consider the function on Q^+, setting $w = \varrho \exp(i\Theta)$,

$$g_\varepsilon(w) = \psi(w) - \frac{2}{\pi} \arg w - \varepsilon \varrho \cos\left(\Theta - \frac{\pi}{4}\right) - \varepsilon \frac{1}{\varrho} \cos\left(\Theta - \frac{\pi}{4}\right) .$$

This is a subharmonic function and we have

$$g_\varepsilon(w) \leqslant 0$$

on the boundary of $Q^+ \cap \{1/R < |w| < R\}$ for R large. We deduce then that

$$g_\varepsilon(w) \leqslant 0 \qquad \forall w \in Q^+ .$$

Since ε is arbitrary we deduce then that

$$\psi(w) \leqslant \frac{2}{\pi} \arg w \qquad \text{on } Q^+$$

i.e.

$$(\alpha) \qquad \varphi(z) \leqslant \frac{2}{\pi} \arg \frac{z+i}{i-z} \qquad \text{on } \mathring{D}^- .$$

With a similar argument one proves that

$$(\beta) \qquad \varphi(z) \leqslant \frac{2}{\pi} \arg \frac{i-z}{z+i} \qquad \text{on } \mathring{D}^+ .$$

From (α) we deduce on $\mathring{D}^-$

$$\arg \frac{z+i}{i-z} = \operatorname{arctg} \frac{2|\operatorname{Re} z|}{1-z\bar{z}} \qquad \text{for } \operatorname{Re} z < 0$$

$$\leqslant 4|\operatorname{Re} z| \qquad \text{for } \operatorname{Re} z < 0 \ \ |z| < \frac{1}{\sqrt{2}}$$

hence

$$\varphi(z) \leqslant \frac{8}{\pi} |\operatorname{Re} z| \qquad \text{on } \mathring{D}^- \cap \left\{|z| < \frac{1}{2}\right\} .$$

Similarly we argue from (β) and we deduce

$$\varphi(z) \leqslant \frac{8}{\pi} |\operatorname{Re} z| \qquad \text{on } D^+ \cap \left\{|z| < \frac{1}{2}\right\} .$$

The last two inequalities prove our statement.

b) Let W be an algebraic cone in $\mathbf{C}^n$ purely dimensional of complex dimension d. Let

$$a \in W_{\mathbf{R}} - \{0\}$$

($W_{\mathbf{R}} = W \cap \mathbf{R}^n$). *We say that W is locally hyperbolic at a if we can find a real linear projection $\pi_a: \mathbf{C}^n \to \mathbf{C}^d$ inducing a proper map of an open neighborhood $U(a)$ of a on W onto an open neighborhood Ω of the origin in $\mathbf{C}^d$ and such that*

$$\pi_a^{-1}(\Omega \cap \mathbf{R}^d) = U(a) \cap W_{\mathbf{R}} .$$

We may assume that Ω is a polycylinder centered at the origin and that $U(a)$ is relatively compact in W so that $U(a) \xrightarrow{\pi} \Omega$ is an unrestricted ramified covering of Ω with finite fibers.

Assume now that W is a purely d-dimensional cone which is locally hyperbolic at any point $a \in W_{\mathbf{R}} - \{0\}$.

We consider the automorphism of W given by the multiplication by i. We have

$$W \cap i\mathbf{R}^n = iW_{\mathbf{R}}$$

and for any point $a \in W \cap i\mathbf{R}^n - \{0\}$ we can find an open neighborhood $U(a)$ of a on W, an open polycilinder Ω in $\mathbf{C}^d$ centered at the origin, a real linear projection $\pi_a: \mathbf{C}^n \to \mathbf{C}^d$ such that (« local hyperbolicity » at $a \in$ $\in iW_{\mathbf{R}} - \{0\}$)

i) $\pi_a | U(a)$ is proper

ii) $\pi_a(U(a)) = \Omega$

iii) $\pi_a^{-1}(\Omega \cap i\mathbf{R}^d) = U(a) \cap i\mathbf{R}^n$.

Let φ be a plurisubharmonic function defined on $U(a)$ and satisfying the conditions

$$(*) \qquad \varphi(\xi) \leqslant |\xi| \quad \text{for } \xi \in U(a)$$
$$\varphi(\xi) \leqslant 0 \quad \text{for } \xi \in U(a) \cap i\mathbf{R}^n .$$

We remark that $U(a) \xrightarrow{\pi} \Omega$ is an unrestricted ramified covering of Ω with a finite number of points in each fiber.

We define on Ω the function

$$\psi(z) = \sup \{\varphi(\xi) | \xi \in U(a), \pi(\xi) = z\} .$$

This function has the properties

$$(\alpha) \qquad \forall z_0 \in \Omega \quad \max \lim_{z \to z_0} \psi(z) = \psi(z_0)$$

(β) there exists a proper analytic subset $\Delta \subsetneq \Omega$ such that at any point $z \in \Omega - \Delta$ ψ is plurisubharmonic.

Indeed it is enough to take for Δ the ramification set of $U(a) \xrightarrow{\pi} \Omega$.

It follows then (by Riemann extension theorem for plurisubharmonic functions) that ψ is plurisubharmonic on Ω.

Because of the properties (*) we deduce that for $c = \sup_{U(a)} |\xi|$ we must have

$$\psi(z) \leqslant c \qquad \forall z \in \Omega$$

$$\psi(z) \leqslant 0 \qquad \forall z \in \Omega \cap i\mathbb{R}^n$$

(this last property because of condition iii) of π mentioned above). By virtue of the previous lemma, with a constant $c' > 0$ depending only from Ω and c (and not from ψ) we must have

$$\psi(z) \leqslant c' |\mathrm{Re}\, z| \qquad \forall z \in \tfrac{1}{2}\Omega\,.$$

Setting $z = \pi_a(\xi)$ we deduce then, setting $U'(a) = \pi_a^{-1}(\frac{1}{2}\Omega) \cap U(a)$

$$\varphi(\xi) \leqslant \psi(\pi_a(\xi)) \leqslant c' |\mathrm{Re}\, \pi_a(\xi)|$$
$$\leqslant c' |\mathrm{Re}\, \xi| \qquad \forall \xi \in U'(a)\,.$$

We have thus proved that

« If W is a purely d-dimensional cone which is "locally hyperbolic" at a point $a \in iW_{\mathbf{R}} - \{0\}$ we can find open neighborhoods

$$U'(a) \subset\subset U(a) \subset\subset W$$

of a in W and a constant $c(a) > 0$ such that for any plurisubharmonic function φ defined in $U(a)$ and satisfying

$$\varphi(\xi) \leqslant |\xi| \qquad \forall \xi \in U(a)$$

$$\varphi(\xi) \leqslant 0 \qquad \forall \xi \in U(a) \cap i\mathbb{R}^n$$

we have also

$$\varphi(\xi) \leqslant c(a) |\mathrm{Re}\, \xi| \qquad \forall \xi \in U'(a)\text{ ».}$$

Now we use the assumption that W is locally hyperbolic at every point $a \in iW_{\mathbf{R}} - \{0\}$.

Set $\Sigma = \{\xi \in W | \xi \in iW_{\mathbf{R}}, |\xi| = 1\}$. This is a compact set and therefore we can find a finite set of points $a_1 \cup \ldots \cup a_k$ in Σ such that

$$\sum \subset \bigcup_{j=1}^{k} U'(a_j) = \Xi .$$

If $c = \sup_{1 \leq j \leq k} c(a_j)$ we derive that:

If φ is a plurisubharmonic function defined on W, if

$$(**) \qquad \begin{aligned} &\varphi(\xi) \leq |\xi| &&\forall \xi \in W \\ &\varphi(\xi) \leq 0 &&\forall \xi \in W \cap i\mathbb{R}^n \end{aligned}$$

then we have also

$$\varphi(\xi) \leq c|\mathrm{Re}\, \xi| \qquad \forall \xi \in \Xi .$$

Let

$$S = (W - \Xi \cap W) \cap \{\xi \in \mathbb{C}^n \big| \, |\xi| = 1\}$$

then S is a compact set and let

$$\varrho = \inf_{\xi \in S} |\mathrm{Re}\, \xi| .$$

By construction we must have $\varrho > 0$ as Ξ is a neighborhood of the region on $W \cap \{|\xi| = 1\}$ where $\mathrm{Re}\, \xi = 0$. We do have therefore

$$|\xi| \leq \frac{1}{\varrho} |\mathrm{Re}\, \xi| \qquad \forall \xi \in S .$$

Hence we must have

$$\varphi(\xi) \leq \max\left(c, \frac{1}{\varrho}\right) |\mathrm{Re}\, \xi| \qquad \text{for } \xi \in W \cap \{\xi \in \mathbb{C}^n \big| \, |\xi| = 1\} .$$

Let now $\xi \in W - \{0\}$ be fixed. We construct the new plurisubharmonic function

$$\psi(\eta) = |\xi|^{-1} \varphi(|\xi| \eta) .$$

If conditions $(**)$ are satisfied by φ we do have also

$$\begin{aligned} &\psi(\eta) \leq |\eta| &&\forall \eta \in W \\ &\psi(\eta) \leq 0 &&\forall \eta \in W \cap i\mathbb{R}^n \end{aligned}$$

therefore for any η with $\eta \in W$ and $|\eta| = 1$ we have

$$\psi(\eta) \leqslant \max\left(c, \frac{1}{\varrho}\right) |\mathrm{Re}\, \eta| \, .$$

We apply this inequality to $\eta = \xi/|\xi|$. We deduce

$$\varphi(\xi) \leqslant \max\left(c, \frac{1}{\varrho}\right) |\mathrm{Re}\, \xi| \, .$$

This shows that *an algebraic cone W pure dimensional and locally hyperbolic at every point of $W_{\mathbb{R}} - \{0\}$ admits the Phragmén-Lindelöf principle.*

We will say that *the given Hilbert complex* (1) *is locally hyperbolic if every irreducible component W of any one of the associated asymptotic cones is locally hyperbolic at every point of $W_{\mathbb{R}} - \{0\}$.* We have thus proved the following

PROPOSITION 28. *If the Hilbert complex* (1) *is locally hyperbolic then $\mathbb{R}^n$ is analytically convex with respect to it i.e.*

$$H^1(\mathbb{R}^n, \mathcal{A}_{A_0}) = 0 \, .$$

SECTION 19

Generalization of a Theorem of De Giorgi and Cattabriga

a) Let a Hilbert complex (1) be given corresponding to a Hilbert resolution (2) of a $\mathcal{P}$-module M ($\mathcal{P} = \mathbb{C}[\xi_1, \ldots, \xi_n]$). Let $\mathfrak{p}_1, \ldots \mathfrak{p}_k$ be the prime ideals of Ass (M) and let V_i denote the algebraic variety in $\mathbb{C}^n$ of zeros of the ideal $\mathfrak{p}_i$ $1 \leqslant i \leqslant k$.

Let

$$V = \bigcup_{i=1}^{k} V_i .$$

The variety V is called the support of M, its irreducible components correspond to the minimal ideals of Ass (M). We will also call V the *naive characteristic variety* of the given complex (1). We have the following

THEOREM 4. *Let a Hilbert complex* (1) *be given and let V denote its naive characteristic variety.*

If

$$\dim_{\mathbb{C}} V \leqslant 1$$

then any open convex set ω in $\mathbb{R}^n$ is analytically convex i.e.

$$H^1(\omega, \mathcal{A}_{A_0}) = 0 .$$

PROOF. α) We recall the Phragmén-Lindelöf principle for $\mathbb{C}$.

Let K be a convex compact set, let $\delta > 0$, let $K_\delta \cap \mathbb{R}^n$ be the δ-neighborhood of K in $\mathbb{R}^n$.

If φ is a subharmonic function on $\mathbb{C}$ such that

$$\varphi(\xi) \leqslant H_K(\xi) + \delta|\xi| \qquad \forall \xi \in \mathbb{C}$$

$$\varphi(\xi) \leqslant 0 \qquad \forall \xi \in i\mathbb{R}$$

then

$$\varphi(\xi) \leqslant H_K(\xi) + \delta|\text{Re}\,\xi| = H_{K_\delta \cap \mathbb{R}^n}(\xi)\,.$$

β) Let $V_1, \ldots, V_k$ be the algebraic varieties associated to the given Hilbert complex (1). For each i we have (because of the assumption)

$$\dim_{\mathbb{C}} V_i \leqslant 1\,.$$

If $\dim_{\mathbb{C}} V_i = 0$ the corresponding asymptotic variety is empty.

If $\dim_{\mathbb{C}} V_i = 1$ then V_i is an irreducible component of V and the corresponding asymptotic variety is a finite set of complex lines issued from the origin.

We have therefore (according to theorem 3 and corollary 1 to proposition 16) to prove that on any complex line L issued from the origin the Phragmén-Lindelöf principle holds for ω.

We distinguish two cases: $L \neq \bar{L}$ and $L = \bar{L}$.

Case $L \neq \bar{L}$: there is only the origin as a real point on L. There exists a constant $c(L) > 0$ such that for $\xi \in L$

$$|\xi| < c|\text{Re}\,\xi|\,.$$

If K is compact convex in ω and $\delta > 0$ is sufficiently small then $K_{c\delta} \cap \mathbb{R}^n \subset \omega$. From $\varphi(\xi) \leqslant H_K(\xi) + \delta|\xi|$ we derive for $\xi \in L$, $\varphi(\xi) \leqslant H_K(\xi) + c\delta|\text{Re}\,\xi| = H_{K_{c\delta} \cap \mathbb{R}^n}(\xi)$. Thus the principle holds on L.

Case $L = \bar{L}$: in this case given K compact convex in ω we choose $\delta > 0$ so small that $K_\delta \cap \mathbb{R}^n \subset \omega$. The Phragmén-Lindelöf principle for $\mathbb{C}$ establishes the principle for L and ω.

REMARK. In particular if $P(D)$ is a differential operator with constant coefficients in $\mathbb{R}^2$, if ω is open convex in $\mathbb{R}^2$ and if $f \in \mathcal{A}(\omega)$ (i.e. is analytic in ω) there exists always a solution $u \in \mathcal{A}(\omega)$ of the equation $P(D)u = f$. This is the case treated by De Giorgi and Cattabriga in [8] for $\omega = \mathbb{R}^2$.

REFERENCES

[1] G. ANDERSSON, *Propagation of analyticity of solutions of partial differential equations with constant coefficients*, Ark. Mat., **8** (1971), pp. 277-302.

[2] A. ANDREOTTI - M. NACINOVICH, *Complexes of partial differential operators*, Ann. Scuola Norm. Sup. Pisa, **3** (1976), pp. 553-621.

[3] A. ANDREOTTI - M. NACINOVICH, *On the envelope of regularity for solutions of homogeneous systems of linear partial differential operators*, Ann. Scuola Norm. Sup. Pisa, **6** (1979), pp. 69-141.

[4] A. ANDREOTTI - M. NACINOVICH, *Analytic convexity*, Ann. Scuola Norm. Sup. Pisa, **7** (1980), pp. 287-372.

[5] N. BOURBAKI, *Algèbre commutative*, Ch. 3 et 4, Hermann, Paris (1961).

[6] E. DE GIORGI, *Solutions analytiques des equations aux derivées partielles à coefficients constants*, *Seminaire Gaulauic-Schwartz*, 1971-72, Exposé 29.

[7] E. DE GIORGI - L. CATTABRIGA, *Una formula di rappresentazione per funzioni analitiche in* $\mathbb{R}^n$, Boll. Un. Mat. Ital., **4** (1971), pp. 1010-1014.

[8] E. DE GIORGI - L. CATTABRIGA, *Una dimostrazione diretta dell'esistenza di soluzioni analitiche nel piano reale*, Boll. Un. Mat. Ital., **4** (1971), pp. 1015-1027.

[9] L. EHRENPREIS, *Fourier analysis in several complex variables*, J. Wiley and Sons, New York, 1970.

[10] H. GRAUERT - R. REMMERT, *Plurisubharmonische funktionen in komplexen rammen*, Math. Z., **65** (1965), pp. 175-194.

[11] H. GRAUERT - R. REMMERT, *Konvexität in der komplexen analysis*, Comment. Math. Helv., **31** (1956), pp. 152-183.

[12] H. GRAUERT - R. REMMERT, *Analytische stellenalgebren*, Springer, Berlin 1971.

[13] A. GROTHENDIECK, *Espaces vectoriels topologiques*, Soc. Mat. de S. Paulo, S. Paulo 1958.

[14] L. HÖRMANDER, *An introduction to complex analysis in several variables*, Van Nostrand, Princeton N. J. 1966.

[15] L. HÖRMANDER, *On the existence of analytic solutions of partial differential equations with constant coefficients*, Invent. Math., **21** (1973), pp. 151-182.

[16] R. NARASIMHAN, *Introduction to the theory of analytic spaces*, Springer lecture notes in Math., n. 25 (1966).

[17] V. P. PALAMODOV, *Linear differential operators with constant coefficients*, Springer, Berlin 1970.

[18] L. PICCININI, *Non surjectivity of the Cauchy-Riemann operator on the space of analytic functions on* $\mathbb{R}^n$. *Generalization to parabolic operators*, Boll. Un. Mat. Ital., **7** (1973), pp. 12-28.

[19] T. MIWA, *On the global existence of real analytic solutions of systems of linear differential equations with constant coefficients*, Proc. Japan Acad., **49** (1973), pp. 500-502.

Pubblicazione parzialmente finanziata dal Consiglio Nazionale delle Ricerche

« Monograf » - Via Collamarini 5 - Bologna
Finito di stampare nell'Aprile 1981